MODERNI KANINIS-HUMAANI SUHDE: KOIRA-IHMINEN –SUHTEESEEN LIITTYVÄN SIDOKSEN MUUTOSVAIKUTUS

Jarmo Saarti[1], Jukka Kananen[1] ja Japo Jussila[2]

[1] Takautuvan logistiikan laitos, UEF
[2] Soveltavan tetrapiloktomian laitos, UEF

*The Association of Expanded Facts of Universal &
Calculated Knowledge (Ass. eFUCK)
Kuopio, 2017*

Julkaisija:
The Association of Expanded Facts of Universal & Calculated Knowledge (Ass. eFUCK)
Kuopio, 2017

Sarja:
Communications in Expanded Facts of Universal & Calculated Knowledge (eFUCK), vol. I.

Toimittajat:
Jarmo Saarti, Jukka Kananen ja Japo Jussila

ISBN: 978-951-568-953-5

© 2017 Saarti, Jarmo / Kananen, Jukka / Jussila, Japo
Kustantaja: BoD - Books on Demand, Helsinki, Suomi
Valmistaja: BoD - Books on Demand, Norderstedt, Saksa

Sisällys

ESIPUHE 5

MODERNI KANINIS-HUMAANI SUHDE: KOIRA-IHMINEN -SUHTEESEEN LIITTYVÄN SIDOKSEN MUUTOSVAIKUTUS 7

1   Johdanto 7

2   Koira-ihminen -sidos 8

2.1 Historiaa 11

2.2   Suora koira-ihminen -sidos 15

       2.2.1 Koira-ihminen -sidoksen määritelmä ja tausta 15

       2.2.2   Määrämittainen koira-ihminen -sidos 17

       2.2.3   Muuttuvamittainen koira-ihminen -sidos 21

       2.2.4   Haaroittuva koira-ihminen -sidos 23

       2.2.5   Koiran kaulalenkki koira-ihminen -sidoksessa 25

       2.2.6 Geneettinen koira-ihminen -sidos 27

3   Koira-ihminen -yhteisö 29

3.1   Sisätila 29

       3.1.1 Koira-ihminen -suhteen koira sohvalla 29

3.2   Ulkotila 31

       3.2.1   Koira-ihminen -suhteen koira tervehtimässä vieraita ohikulkijoita 31

       3.2.2   Koira-ihminen -suhteen koira aloitteen tekijänä tervehtimisessä 33

3.2.3 Koira-ihminen -suhteen ihminen koiran jätösten muodostumisen seuraajana..................................37

3.2.4 Ihmisten välisten keskustelujen aloitus koira-ihminen -suhteeseen liittyvässä ulkoilutilanteessa..........39

4 Koira-ihminen -suhdemuutunnan harhaisuuksia ja niiden käytännön seurauksia......................................43

4.1 Koira-ihminen -suhteen koiran hämmentävä rooli sisätiloissa..43

4.2 Koira-ihminen -suhteen koira tekosyynä autokaupassa..44

4.3. Koira-ihminen -suhteen oikeustoimikoira..............46

4.4. Koira-ihminen -suhteen koira on ihmistä viisaampi..48

ESIPUHE

Kommentit Mäntyniemen koirakopista, josta kuuluu tyytyväisyys koira-ihminen -suhteen koiran ällistyttävän tarkasta valinnasta. Se pohjustaa tietä toiselle presidenttikaudelle. Vaalikirja, joka tasoittaa presidenttiparin vaiteliaamman osapuolen, koira-ihminen -suhteen koiran, tien jatkoon jo ensimmäisellä kierroksella. Susi olisi ollut asioiden laidasta ihmeissään, jos se olisi osannut nähdä tulevaisuutensa koira-ihminen -suhteen aamuhämärissä, nuotiota lähestyessään ja harkitessaan koira-ihminen -suhteen ihmisen kesyttämistapoja. Kuinka korkealle hänen yritelmänsä tulevat johtamaan ja kuinka vallitsevaksi koira-ihminen -suhde kehittyykään?

Koirakirja niille, jotka ovat koirakirjansa ansainneet. Turha luulla, että koira-ihminen -suhteen arvoitus, monimuotoisuus tai edes tarkoitus selviäisi kattavassakaan tarkastelussa, joten olemme vaatimattomasti päättäneet ratkaista sen lisäksi myös olemassaolon tarkoituksen. Monthypythonilainen logiikka lähestyy näitä kysymyksiä absurdeista lähtökohdista, me otimme tämän lähestymiskulman huomioon, reivasimme sen reilusti tarkoituksenmukaisen tieteellisen lähestymistavan suuntaan ja teimme mitä koira-ihminen -suhteen ihmisen pitää tehdä.

Kuopiossa, neliulotteisessa avaruudessa, aikajatkumon kärjessä
6.12.2017

Jarmo Saarti   Jukka Kananen   Japo Jussila

MODERNI KANINIS-HUMAANI SUHDE: KOIRA-IHMINEN –SUHTEESEEN LIITTYVÄN SIDOKSEN MUUTOS-VAIKUTUS

Jarmo Saarti[1], Jukka Kananen[1] ja Japo Jussila[2]

[1] Takautuvan logistiikan laitos, UEF; [2] Soveltavan tetrapiloktomian laitos, UEF

1 Johdanto

Koira-ihminen -suhteen (määritelmä: henkinen vuorovaikutus, jossa on myös suoria fyysisiä ulottuvuuksia) kehittymistä on pohdittu runsaasti, erityisesti historiallisessa perspektiivissä ja sosiobiologisissa tutkimuksissa. Koirasta aktiivisen roolin hakijana, jopa määräävänä toimijana, koira-ihminen -suhteeseen liittyvässä yhteisössä ei liene epäselvyyttä, vaikka roolin suhde koiraan on joissakin, lähinnä viimeaikaisissa poikkitieteellisissä tutkimussuuntauksissa (vrt. feministinen tutkimus) tutkimuksissa asetettu kyseenalaiseksi.

Tässä tutkielmassa pohditaan yhtä koira-ihminen -suhteen aluetta, nimittäin koira-ihminen -suhteeseen liittyvän *koira-ihminen -sidoksen* (määritelmä: ihmisen koiraan yhdistävä fyysinen, aina kiinteäksi muovautuva sidos) yleistä ja viimeaikaista muuntumista ja sen seurauksia koira-ihminen -suhteen muuntumiseen ja roolittamiseen.

Lähtökohtana on sekä biologisesta todellisuudesta irrotettu, osin vaivihkaisen biologissidonnainen, fenomenologisen muuntumisen teoria, että koiran ja ihmisten käyttäytymistä ja ihmisyyttä koskevat sosiologiset teoriat.

Aiheen ajankohtaisuudesta voinee käyttää esimerkkinä marraskuussa 2017 julkaistua löytöä koira-ihminen -suhteen alkuvaiheesta eräässä saudiarabialaisessa taideteoksessa, jossa oli kuvattuna alkuvaiheen koira-ihminen -sidos koira-ihminen -suhteeseen liittyvässä yhteistoimintakuvauksessa. Teos on ajoitettu noin 9000 vuotta e.a.a. ja kertonee omalla karulla tavallaan koiran merkitsevästä roolista koira-ihminen -suhteessa, jollei kyseinen taideteos ja koko löydös myöhemmin paljastu tieteelliseksi manipulaatioksi.

2 Koira-ihminen –sidos

Tässä kappaleessa esitellään koira-ihminen -sidoksen periaatetta ja muuntumista. Tämän kirjoituksen laajuuteen ei katsottu tarpeelliseksi sisällyttää koira-ihminen -sidoksen koiranpuoleisen kiinnittymiskohdan luonnetta ja muuntumista, muutoin kuin sellaisten kehitysvaiheiden osalta, joilla on erityinen asema koira-ihminen -sidoksen merkityksen, luonteen ja muuntumisen ymmärtämiseksi. Koira-ihminen -sidoksella tarkoitetaan tässä yhteydessä ohutta fyysistä, määrämittaista tai venyvää, useimmiten kiinteää, yhteyttä koiran ja ihmisen välillä, jolle on annettu useita käytännön tilannetta kuvaavia nimityksiä, kuten 'remmi', 'talutusnuora', 'naru', jne. (Kuvat 1 ja 2).

Kuva 1. *Koira-ihminen -sidoksen karkea periaate on yhdistää koira ihmiseen pysyväisluonteisella, kiinteällä sidoksella. Koira-ihminen -sidos on merkitty nuolella. Koira on hahmoista pienempi, ihminen suurempi. Kuvan mittasuhteet ovat lähellä yhtä perinteisistä, jopa klassisista, ja siten myös tavanomaisinta koira-ihminen -suhteen mittasuhteista. Kuva CC BY 4.0.*

Siirrymme lyhyen historiakatsauksen jälkeen suoraan ns. vahvan koira-ihminen -sidoksen vaiheen kuvaukseen ja siihen liittyviin pohdintoihin. Heikon koira-ihminen -sidoksen vaihetta on tarkasteltu valitsemaamme aiheeseen liittyvissä ja sitä sivuavissa töissä, kuten susi- ja kojoottitutkimuksissa. Toivomme, että asiasta kiinnostuneet hakevat lisätietoa esimerkiksi paikallisesta kirjastosta.

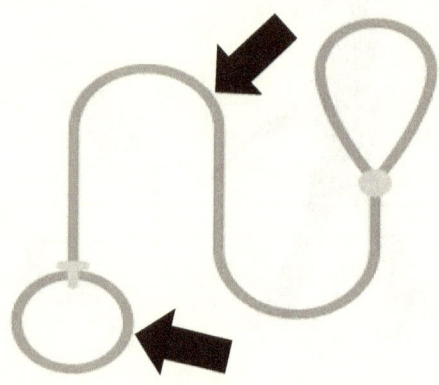

Kuva 2. *Periaatekuva esimerkkinä täydellisesti kehittyneen koira-ihminen -sidoksesta. Koira-ihminen -sidoksen varsinainen sidososa merkitty ylemmällä nuolella (vrt. Kuva 1), jonka päässä on tyypillinen kiinteä silmukka koira-ihminen -sidoksen ihmispuolen kiinnityksen varmentamiseksi. Huomatkaa koira-ihminen -sidoksen koiranpuoleinen kiinnityslenkki (pyöreä muoto, alempi nuoli), jonka luonnetta ei erikseen käsitellä tässä tutkielmassa. Kuva CC BY 4.0.*

2.1 Historiaa

Koira-ihminen -sidoksen ensimmäiset, toistaiseksi dokumentoidusti löydetyt ja kuvatut merkit on ajoitettu noin 30,000 vuotta vanhoiksi. Koira-ihminen -suhde on siis vanhempi kuin maanviljelyksen kehittämisen historia ja siten vanhempi myös kuin maanviljelyyn ja hajasijoittamiseen keskittynyt poliittinen liikehdintä, joka yleisessä poliittisessa yhteydessä tunnetaan termillä 'kipuilu' tai 'siltarumpuilu' (jälkimmäistä ei pidä sekoittaa jo koira-ihminen siteen alkuaikojen heimorumpuiluun).

Ihminen liittyi koiran seuraan alkujaan Euroopassa, ei Afrikassa, kuten on väitetty, koska Afrikka ei ole ollut harmaasusien alkuperäistä levinneisyysaluetta. Nykyisin koirina (vrt. vartiokoira, lemmikkikoira, metsästyskoira ja hirvikoira) tunnetut eliöt ovat geneettisten tutkimusten mukaan läheisintä sukua muinaisille eurooppalaisille susille.

Koiria hyödyttänyt rinnakkaiselo alkoi, kun alkukoirat siirtyivät elämään ihmisasutusten läheisyyteen pysyvästi aiempien lyhyiden vierailujen asemesta. Koirat oppivat arvostamaan etuja, joita suurien ja vaarallisten lihansyöjien kesyttäminen tarjosi erityisesti helpottamalla ravinnon hankkimista: koiran tavoitteena oli kesyttämällä ja kotouttamalla ihminen välttää rasittavaa ja laumaa kuluttavaa metsästystä ja jättää se ihmisen tehtäväksi.

Vähitellen tästä kanssakäymisestä muodostui koira-ihminen -suhde. Myöhempi koira-ihminen -suhteen historia, erityisesti nykymuotoinen koira-ihminen -suhde, osoittanee tämän päämäärän onnistumisen asteen. On esitetty nimenomaan, että koira kesytti ensin naisen, jota näkemystä on käsitelty mm. lapsettomuusteoriaa tai hoivavietin siirtymä -teoriaa käsittelevissä pohdinnoissa ja radikaaleissa seksuaalisuuden sublimoinnin teorioissa (Kuva 3). Koiran suhdemuutuntaa koira-ihminen -suhteen ihmisyhteisön osana kuvannee esimerkiksi se, että arkeologisten löydösten mukaan ihmisiä on haudattu koirien kanssa jopa 12,000 vuotta sitten, ehkä jo aikaisemminkin. Koiria on kuvattu maalauksiin ja veistoksiin noin 9,000 vuotta sitten, joka korostaa koira-ihminen -suhteen merkitystä. Tosin tässä yhteydessä voisi puhua hautausrituaalin antaman kuvan vahvistamisesta.

Varhaisimmat kirjalliset tiedot koira-ihminen -suhteen koiran monista toimista koira-ihminen -suhteen ihmisten parissa ovat peräisin muinaisesta Lähi-idästä ja Välimeren alueelta, Kaksoisvirtain maasta ja Egyptistä yli neljän tuhannen vuoden takaa.

Koiran merkitystä koira-ihminen -suhteen kulmakivenä korostavat useat kansantarinat ja myytit, joiden mukaan ihminen ei olekaan, ainakaan suoraan, polveutunut apinasta. Tässä vaiheessa on hyvä huomata tämän merkitys sille keskustelulle, jota on käyty 'puuttuvan lenkin' tai 'välimuodon' ongelmasta ihmisen polveutumis- ja kehityskaaressa.

Myyttien mukaan Aasian kansoista mm. kirgiisit, kazakit, korealaiset, mongolit, tiibetiläiset, tunguusit, turkkilaiset, uiguurit ja ainut polveutuvat koirista. Pohjois-Amerikan arktisilla alueilla inuiitit ovat pitäneet koiria kantaisinään. Tätä perinnettä ovat myös ihmissusiin liittyvät tarinat ja myytit sekä kettumytologia, joka osoittaa selviä älykkyyteen liitettyjä ominaisuuksia suhteen dominantissa osapuolessa.

Kuva 3. Koiran rooli koira-ihminen -suhteessa voi tietyissä yhteiskunnallis-biologisissa tilanteissa muodostua, erityisesti hoivavietin paineesta (vrt. lapsettomuusteoriat), muistuttamaan koira-ihminen -suhteen ihmisen jälkikasvun korvikeroolia. Kuvan esimerkissä voi nähdä viitteitä tällaisesta asetelmasta. On kuitenkin huomautettava, että esimerkkikuvan perusteella ei tule vetää liian pitkälle meneviä johtopäätöksiä tämän kaltaisen koira-ihminen -suhteen kehittymiseen liittyvistä sukupuolipainotuksista (tai yleensäkään sukupuolittumisesta) koiran tai ihmisen osalta. Kuva CC BY 4.0.

Tämän koira-ihminen -suhteen erityispiirteen, ihminen toimimassa tiedostamattaan koiran palvelijana koira-ihminen -suhteessa, on myös väitetty edistäneen koiran asemaa parantaneiden vuorovaikutusmekanismien motiivien hämärtymisen koira-ihminen -suhteen ihmisen osalta (klassinen 'hämäysteorian' mukainen hämäystoimi).

Kieli- ja käyttäytymistieteitä viime aikoina yhdistelleet koulukunnat ovatkin pohtineet tämän, 'liiviin uimiseksi' kuvatun, koira-ihminen -suhteen varhaisvaiheen merkitystä siihen, että koira-ihminen -suhteen ihmiset alkoivat pukea koiria talvisin (muulloinkin kylmien ilmojen sattuessa) erityisiin liiveihin. Onpa nähty tätäkin pidemmälle vietyjä pukeutumisrituaaleja.

Esimerkkinä koira-ihminen -suhteen kehittymisestä yleisellä tasolla on parsilaisten pyhä kirjan Avestan kuvaukset koiran suhteesta ihmiseen (huomatkaa kirjaimellinen määritelmä: koira-ihminen -suhde!). Kuvaukset ajoittuvat todennäköisesti Zarathustran aikoihin (1000–600 e.a.a.) ja tekstissä verrataan koiran luonnetta kahdeksaan erilaiseen ihmiseen.

Koira on kuin pappi, koska se tyytyy vähään, on kärsivällinen ja syö tähteitä. Se on kuin soturi, sillä se on urhoollinen taistelija. Toisaalta koira on kuin aviomies, joka valvoo ja vartioi ja nukkuu kevyesti vaaran takia. Se on myös kuin kiertävä laulaja, sillä se on innokas laulamaan. Koska koira on häpeämätön syömisessään, pitää hämärästä ja vaanii saalista, se vertautuu varkaaseen ja villi-ihmiseen. Kurtisaanin tavoin se pitää laulamisesta, juoksentelee kaduilla ja on huonosti kasvatettu ja ailahteleva. Kaiken kukkuraksi koira on kuin lapsi, sillä se on hellä, rakastaa nukkumista, pyrkii puhumaan koko ajan ja kaivaa maata käpälillään.

14

2.2 Suora koira-ihminen –sidos

2.2.1 Koira-ihminen –sidoksen määritelmä ja tausta

Koira-ihminen -sidosta on pidetty merkittävänä ihmisyhteisön kehitystä ohjaavana biofyysisenä rakenteena. Itse koira-ihminen -sidos on pelkistetty kytkös koiran ja ihmisen välillä, usein vain ohut luonnon- ja tai tekokuidusta valmistettu sidos, naru tai köysi, tyypillisesti ohuen narumainen ja pituudeltaan sellainen, että ihmisen etäisyys koirasta on alle 2 m (Kuva 4). Sidokselle on ominaista, että sidoksen yhdistämien eliöiden välille (koira-ihminen -sidoksessa koiran ja ihmisen välille) muodostuu kiinteä, valtasuhteiden käyttöön soveltuva kytkös.

Koira-ihminen -sidos sitoo ihmisen kiinteästi koiraan ja voi täten helposti, joskaan ei säännönmukaisesti, siirtää johtajan roolia koiralle koira-ihminen -suhteessa. Koira-ihminen -sidoksen pitkä historia ja mielikuvituksellinen kehityskaari sekä sidoksen ulkonäön, että toiminnan osalta osoittaa, että koira-ihminen -sidos on ollut merkittävässä roolissa koira-ihminen -suhteen evoluutiossa heti koiran kesytettyä ihmisen.

Voidaan väittää, joskin tähän väitteeseen on usein liitetty lukuisia varaumia, että koira-ihminen -sidos on säilynyt huomattavan pitkään yhtenä modernin yhteiskunnan tukipilareista. Koira-ihminen -sidoksen rooli yhteiskunnallisena vaikuttajana ja suunnannäyttäjänä on ainutlaatuista tämänkaltaiselle ohuelle sidokselle, josta on biologiassa käytetty nimitystä 'flagella'.

Vastaavia sidokseen pohjautuvia isäntä-parasiitti -suhteita on kuvattu myös muiden eliöiden osalta (esimerkkinä unitaudin aiheuttaja), ja kiinteän sidoksen tai fyysisen suhteen on todettu merkittävästi vaikuttavan isäntä-parasiitti -suhteen vuorovaikutusmekanismeihin ja isäntä-parasiitti -suhteen eliöiden muuntumiseen. Tämän esimerkin innoittamana on syytä todeta, että erilaiset isäntä-parasiitti -suhteet, jotka perustuvat kiinteän rakenteen, kuten sidoksen, muodostumiseen isäntä-parasiitti -suhteen eliöiden välille (vrt. koira-ihminen -sidos), voivat olla merkittäviä ihmisyhteisöjen ja ihmiskunnan kannalta.

Kuva 4. *Koira-ihminen -sidoksen periaate (koira-ihminen -sidos merkitty nuolella). Koira on esitetty harmaana, koira-ihminen -sidoksen koiranpuoleinen pää pilkottaa koiran karvoituksen seasta oikeanpuoleisen korvan alla. Kuva on pelkistys, jonka kehitystä ja erikoistapauksia käsittelemme laajemmin itse tekstiosuudessa. Kuva CC BY 4.0.*

2.2.2 Määrämittainen koira-ihminen –sidos

Koiran ensimmäinen askel tasa-arvoisuuteen ja mahdollisesti jopa hallintaan suhteessaan ihmiseen oli *määrämittaisen koira-ihminen -sidoksen* kehittäminen ja käyttöönotto koira-ihminen -suhteen elimellisenä osana (Kuva 5). Myöhemmässä vaiheessa koira-ihminen -sidoksen muuntuminen ja sen eri muodot ovat merkittävästi vaikuttaneet koira-ihminen -suhteen yhteisöllisiin kehitysmuotoihin.

Koira-ihminen -suhteen koiran kiinteä suhde koira-ihminen -suhteen ihmiseen muutti ratkaisevasti aiempaa, vapaamuotoista yhteyttä koiran ja ihmisen välillä. Koira-ihminen -sidokseton ihminen oli vapaa kulkemaan yksin ja tutkimaan ympäristöään omien tarpeidensa mukaan ja tyydyttämiseksi ennen kiinteän koira-ihminen -sidoksen kehittymistä (Kuva 6).

Sidoksen kehittymisen jälkeen on koiran rooli korostunut koira-ihminen -suhteessa. Koira-ihminen -parin liikkuessa pesän, suojan, tai nykyisin kodin, ulkopuolella, on ihminen aiempaa enemmän kiinnittänyt huomionsa ympäristöön ja sen tapahtumiin koiran ohjaamana ja menettänyt siten merkittävän osan omaa tahtoaan.

Koira-ihminen -sidoksen koiranpuolinen pää kehittyi tiukasti koiran kaulan ympärille sopivaksi lenkiksi koira-ihminen -sidoksen kehityksen ensimmäisessä, varhaisessa vaiheessa. Koira-ihminen -sidoksen varhaisimmissa kehitysvaiheissa lienee ollut edullista koiran roolin vaivihkaisen kehittymisen kannalta, että koira-ihminen -sidoksen koiranpuolisen kiinnittymiskohdan on voinut olettaa ilmentävän ihmisen määräävää roolia koira-ihminen -suhteen kehityksessä.

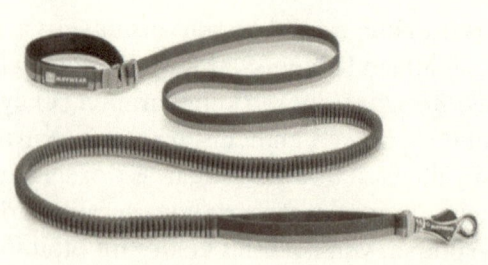

Kuva 5. *Kiinteän koira-ihminen -sidoksen periaatemalli. Tästä, näennäisesti pelkistetystä mallista on nopeasti kehittynyt paikallisesti ja ajallisesti sopeutuneita muunnelmia, joiden voi olettaa toimineen sekä koiran että ihmisen kannalta edullisesti koira-ihminen -suhteessa. Sidoksen evolutiivisestä kehityskaaresta on vielä niukasti tietoa, mutta koiran määräävän aseman lisääntyessä koira-ihminen -sidoksen tutkimus, muun asiaan liittyvän tutkimuksen ohella, saanee merkittävästi lisää rahoitusta. Kuva CC BY 4.0.*

Alkujaan määrittelemättömän, yleisemmän sidoksen vakiintumiselle koira-ihminen -sidokseksi on ollut edullista se, että ihminen on hyväksynyt koira-ihminen -sidoksen omalta kannaltaan edullisena kehitysvaiheena koira-ihminen -suhteessa. Kehitysvaiheen käynnistyminen on vallitsevien teorioiden mukaan ihmisen tahdosta riippumaton kehityskokonaisuus, eikä ole vaatinut ihmisen aktiivista roolia koira-ihminen -suhteen alkuvaiheessa (teoria on kiistanalainen, mutta vallitseva).

Tämä vaivihkaisuus on ollut oletusten mukaan tärkeää ennen tunnepohjaiseen sitoutumiseen kehittymisen astetta.

Erään, moneen kertaan kyseenalaistetun, teorian mukaan ihmisen määräävä rooli koira-ihminen -suhteen yleisen kehittymisen alkuvaiheessa on ollut monessa yksittäistapauksessa vaarantamassa koira-ihminen -suhteen kehittymisen nykyisenkaltaiseen suuntaan.

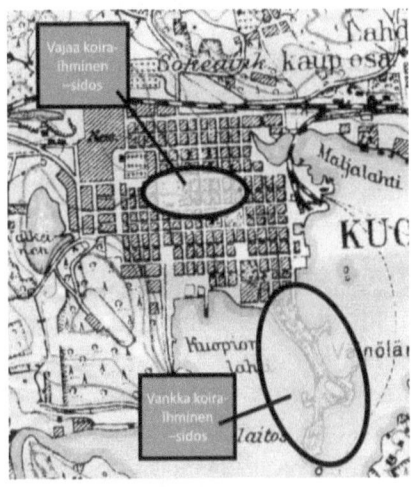

Kuva 6. *Kiinteän koira-ihminen -sidoksen vaikutus ihmisen liikkumiseen kaupunkiympäristössä, esimerkkinä nimeltä mainitsematon kaupunki Savosta. Laaja savolainen liikunta-aineisto käsitelty KuumaPiste™ analyysillä, joka tuottaa 98 % tarkkuuden liikkumisalueiden ydinalueen sijaintiin ja sen muutoksiin sekä antaa luotettavuutta kuvaavia muuttuja-arvoja (p*** < 0.0001 tässä aineistossa). KuumaPiste™ analyysin kuvatulosteesta näkyy selvästi koira-ihminen -sidoksen vaikutus erityyppisten koira-ihminen -suhteen ihmisten liikkumisalueen siirtymään, esimerkissämme ihmiskontaktialueelta koiran kannalta herkullisemmalle seudulle (vrt. ympyröidyt KuumaPiste™ alueet). Kuva CC BY 4.0.*

Koira on epäonnistuneen koira-ihminen-suhteen alkuvaiheessa saatettu siten nähdä hyötyeläimenä, kuten ruokana, perheenjäsenen asemesta. Tämä on ollut omiaan vaarantamaan koira-ihminen -suhteen pitkäaikaisen kehittymisen yksilötasolla ja samalla ollut esteenä nykyisenkaltaisen koira-ihminen -sidoksen kehittymiselle.

Alkuperäisistä koira-ihminen -sidoksista omituisimmat liittyvät tarinaperinteeseen, jossa vanhempansa menettäneet koiranpennut ovat adoptoineet itselleen ihmisvanhemmat. Tähän liittyy koira-ihminen -sidoksen kehittyminen meidän aikoinamme kohti täydellistä syrjäyttämistä eli tilannetta, jossa koiranpentu syrjäyttää täysin ihmislapsen kaupunkilaistuneissa koira-ihminen -suhteeseen liittyvissä perheyhteisöissä.

On epäselvää, liittyikö koira-ihminen -sidoksen koiran ihmiseen liittävä osa kiinteäksi koiran kaulalenkin yhteyteen ensin vai oliko koira-ihminen -sidoksen kehityssuunnassa heti alusta sekä kiinteästi kaulalenkkiin liittyvä jatke, että erillinen, irtoava jatke olemassa jo tässä kehitysvaiheessa. Selvää kuitenkin on, että koiran kaulalenkistä irrotettava jatke syrjäytti kiinteän jatkeen kohtuullisen nopeasti.

Koira-ihminen -sidoksen koiran kaulalenkin merkitystä ja sitä, miksi koira hyväksyi pysyvän kaulalenkin, on pohdittu useasta eri lähtökohdasta. Emme tässä puutu asiaan syvällisemmin, mutta mainitsemme kuitenkin, että koira-ihminen -sidokseen liittyvä koiran pysyvä kaulalenkki on ilmeinen omistus- ja hallintasuhteen merkki ja voisi olla mielenkiintoista pohtia, miksi koira hyväksyi tämän alistussuhteen merkin, *kaulasidoksen*, koira-ihminen -suhteen alkuvaiheessa. Tässä on selkeä eläinpsykologinen tutkimussuunta, joka voisi merkittävästi auttaa ymmärtämään nykyisen koira-ihminen -suhteen luonnetta.

On myös esitetty, että koira-ihminen -sidoksen näennäisesti koiraa alistavat piirteet olivat harkittuja ja koiran kannalta edullisia koira-ihminen -suhteen hallitsemisen tasapainoon liittyvien osatekijöiden kehittymiseksi koira-ihminen -suhteen koiran kannalta myönteiseen suuntaan. On esitetty, että koira-ihminen -sidoksen luonne, joka hämäsi ihmistä, antoi koiralle vaivihkaisen tien määräävään asemaan, tosin tämä 'kohdennetun evolutiivisen kehittymisen määrääminen' on kyseenalaistettu useasti ja teoriaa tukevia tutkimustuloksia on vähänlaisesti.

Samoin on esitetty, että koiran kiristyvä kaulalenkki, joka liittyy kiinteästi koira-ihminen -sidoksen koiran ihmiseen liittävään osaan, olisi ollut yhtenä tärkeänä varhaisen kehitysvaiheen osana koira-ihminen -sidoksen kehityksessä. Tätä näkemystä puoltaa mm. ihmisen pyrkimys kesyttää koira jo varhaisessa vaiheessa ja rangaistuksenomainen koiran kaulalenkki suopeiden ominaisuuksien valinnan ohella on voinut auttaa tämän päämäärän saavuttamisessa. Edellisessä kappaleessa esitetyn lisäksi, tämä, jopa masokistis-sadistiseksi luonnehdittava, ulottuvuus koira-ihminen -suhteessa on mielenkiintoinen.

2.2.3 Muuttuvamittainen koira-ihminen –sidos

Koiran kannalta ratkaiseva koira-ihminen -sidoksen kehityshyppäys oli *muuttuvamittaisen koira-ihminen -sidoksen* kehittyminen (Kuva 7). Tämä antoi koiralle lisää vapautta koira-ihminen -suhteessa ja vaati ihmiseltä entistä enemmän resursseja koiran tarkkailuun, joten muut ihmisen ke-

21

hittymisen kannalta oleelliset oheistoiminnot, kuten parittelutilanteisiin hakeutuminen, menettivät ratkaisevasti resursseja.

Kuva 7. *Muuttuvamittaisen koira-ihminen -sidoksen yksinkertaistettu malli, joka sallii koira-ihminen -suhteen koiralle merkittävän liikkumisvapauden verrattuna määrämittaiseen sidokseen. Tätä evolutiivista hyppäystä on verrattu polkupyörän merkitykseen ihmisen geneettisen isolaation purkajana suomalaisella maaseudulla. Huomaa kehittynyt koira-ihminen -sidoksen ihmisenpuolinen osa, johon on (ehkän mutaation vuoksi) kehittynyt kuntoiluvälinettä muistuttava kahvaosa. Tämän kahvaosan merkityksestä ollaan erimielisiä, joskin yleisesti hyväksyttynä, johtava teoriana on koira-ihminen -suhteen ihmisten taipumus rituaaliliikuntaan ja siten kiinnostus tämänkaltaisesti muotoiltujen tuotteiden käsittelyyn. Kuva CC BY 4.0.*

Muuttuvamittainen koira-ihminen –sidos antoi koiralle mahdollisuuden toteuttaa ihmisen ohjaamista entistä joustavammin ja yllättävämmin ja jopa ohjata ihmisen hakeutumista parittelutilanteisiin. Onkin mahdollista, että koira on tämän koira-ihminen –sidoksen kehitysvaiheen ansiosta ottanut aiempaa enemmän vastuuta ihmispopulaation geneettisestä suuntaamisesta.

Koira-ihminen –sidoksen kehittyminen muuttuvamittaiseksi aiheutti sen, että ihmisen mahdollisuus kontrolloida koiraa väheni. Muuttuvamittaista koira-ihminen –sidosta voisi verrata polkupyörän keksimiseen ja käyttöönottoon, jonka vuoksi ihmisen geneettinen isolaatio murtui. Muuttuvamittainen koira-ihminen –sidos puolestaan antoi koiralle, ulkoiselle tekijälle, mahdollisuuden geneettisen valinnan ohjaamiseen.

On myös huomattava, että koiran kannalta sen omien parittelutilanteiden todennäköisyys toki hieman kasvoi muuttuvamittaisen koira-ihminen –sidoksen ansiosta, mutta tässä tapauksessa on kyseessä kuitenkin valtasuhteiden suuri muutos ja koiran hallintaosuuden lisääntyminen. Muuttuvamittaisen koira-ihminen –sidoksen hallinta on ihmisen ja koiran valtataistelun osana erityisen tärkeä osatekijä, kun koira ulkoilee.

2.2.4 Haaroittuva koira-ihminen –sidos

Koira-ihminen –sidoksen mielenkiintoinen muuntoversio on kahden kiinteän kappaleen välille kehittynyt sidos, johon on kiinnittynyt siitä eriävä sidos, jonka toinen pää on kiinni koirassa (vrt. Kuvat 2, 5 ja 7). On selvää, että tämä *haaroittuva koira-ihminen –sidos* on yksi eriytynyt koira-ihminen –sidoksen kehitysmuoto, mutta on epäselvää, missä

23

vaiheessa tämä haaroittuva koira-ihminen -sidos on eriytynyt omaksi, erilliseksi sidosmuodokseen.

On myös mielenkiintoista, että tämän sidoksen yksi osa ei ole yhteydessä elävään kudokseen (jos oletetaan, että haaroittuvan koira-ihminen -sidoksen vaakaosan kiinnityskohdat eivät ole elävää kudosta, vrt. Kuva 8) ja tähän sidokseen on liittynyt koirasta lähtevä sidos. Lisäksi haaroittuvassa koira-ihminen -sidoksessa ihmisen kiinnittyminen sidokseen on satunnaista ja pääsääntöisesti erittäin lyhytaikaista. Tutkimukset tämän haaroittuvan koira-ihminen -sidoksen erityispiirteen vaikutuksesta koira-ihminen -suhteen ovat kesken ja, kuten usein koira-ihminen -suhteen tutkimuksessa, esitetyt teoreettiset näkemykset ristiriitaisia ja kiistanalaisia.

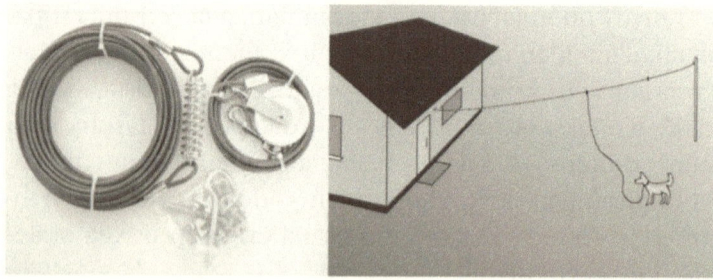

Kuva 8. *Haaroittuva koira-ihminen -sidos yksittäisiksi osiksi purettuna (vasemmalla) ja esimerkki haaroittuvan koira-ihminen -sidoksen käytännön sovelluksesta (oikealla). Haaroittuva koira-ihminen -sidos on rakenteeltaan monimutkainen kehitysvaihe, kuten vasemmalla oleva räjähdyskuvamainen esitystapa kertoo. Kuva CC BY 4.0.*

Haaroittuvan koira-ihminen -sidoksen yksi sidososa (vaakaosa) on perinteisesti ollut kiinni eloperäisessä materiaalissa, kuten puisessa rakenteessa, joka on voinut muodostaa ihmisen asuttaman rakennuksen seinän. Yleinen haaroittuvan sidoksen kiinnittymisalusta on perinteisesti ollut myös elävä puu (mm. mänty on suosittu, samoin muutamat pihapiirin koristeelliset lehtipuut), tosin kaupungistumisen myötä elävään puuhun kiinnittyneet haaroittuvat koira-ihminen -sidokset ovat harvinaistuneet. Tähän lienee syynä kaupunkimaiselle alueelle tyypillinen elävien puiden vähäisyys ja haaroittuvan koira-ihminen -sidoksen kannalta puiden epäsuotuisa esiintyminen.

2.2.5 Koiran kaulalenkki koira-ihminen -sidoksessa

Käsittelemme tässä lyhyesti koira-ihminen -sidoksen koiranpuoleista lenkkiä (*kaulalenkki*), vaikka olemme useaan otteeseen todenneet, että emme käsittele tätä asiaa. Pitäydymme yhä kannassamme, että emme käsittele tätä asiaa, vaan sivuamme sitä tämän tutkielman valmistelun aikana saamamme yleisöpalautteen vuoksi. Emme siis varsinaisesti käsittele tätä asiaa tässä yhteydessä, mutta sivuamme sitä. Pyynnöstä.

Koira-ihminen -sidoksen oleellinen osa on koiran kaulalenkki, joka on kiinni erityyppisillä pienrakenteilla koira-ihminen -sidoksen koiranpuoleisessa päässä. Koiran kaulalenkki on, nimensä mukaisesti, koiran kaulan ympärillä ja tavanomaisesti niin tiukka, että se ei pääse liukumaan koiran kallon yli.

Tällä on merkitystä sekä koiran aseman, että koira-ihminen -vuorovaikutussuhteen kehittymisessä ja vakiintumisessa.

Koiran kaulalenkillä on merkityksensä myös koiran sekä ihmisen määräävän aseman kehittymisen osalta osana koira-ihminen -suhdetta.

Yleisesti oletetaan, että koiran kaulalenkki kehittyi koira-ihminen -sidoksen kehittymisen alkuvaiheen välittömänä seurauksena. Koiran kaulalenkki on koira-ihminen -suhteen merkillinen erityistapaus, jossa ihminen pyrkii hallitsemaan koiraa.

On menty jopa niin pitkälle, että modernin yhteiskunnan lainsäädännössä on koira-ihminen -sidoksen erityispiirre, koiran kaulalenkki, mainittu koira-ihminen -suhteen ohjailussa. Useilla kaupungistuneilla alueilla koiran tulee olla aina koiran kaulalenkin avulla koira-ihminen -sidokseen kytkettynä (erikoistapauksia lukuun ottamatta) ja jopa maaseudulla on määräyksiä koiran kaulalenkin sekä koira-ihminen -sidoksen käytöstä luonnonvaraisten eläinten elämänrytmin mukaan määriteltyinä ajankohtina.

Tämän ihmisyhteisön lainsäädäntöön otettu, koira-ihminen -sidoksen evoluutioon liittyvä määräys kertoo harmillisen selkeästi modernin ihmisen hämärtyneestä käsityksestä ihmisen kyvystä rajata ja ohjailla evolutiivisia prosesseja. On toki myös mahdollista, että tämänkaltaisilla määräyksillä on evolutiivista painoarvoa, tosin nykyinen käsitys lainsäädännön painosta evoluution suuntaajana on vähintäänkin epäilevä, jopa epäpätevä.

Koira-ihminen -sidoksen rajoittama koira voi osin hallita ja samalla huomattavasti rauhoittaa koiraa pelkääviä ihmisiä, on tosin epäselvää, mitä erityistä koira saavuttaa tällä koira-ihminen -suhteen osalta. On myös havaittu, että koira voi käyttää kaulalenkkiä manipuloidakseen ihmistä ja saadakseen sen avulla valta-aseman koira-ihminen -suhteessa.

Käytännön tilanteita tarkkailtaessa on kokeissa havaittu, että koira voi saavuttaa merkittäviä etuja valittamalla koira-ihminen -sidoksen kaulalenkistä, vaikka siitä ei ulkopuolisen havainnoitsijan ole mahdollista nähdä olevan haittaa koiran normaaleille toiminnoille.

Tässä tilanteessa koira voi siten sekä hämärtää ihmisen käsitystä oikeasta ja väärästä, manipuloida arjen moraalikäsityksiä, ja samalla saada etuja ihmisen tuntiessa alemmuutta aiheuttaessaan vaivaa tai tuskaa ohjaajakseen, ystäväkseen tai johtajakseen mieltämälleen eliölle, koiralle. Koira voi siten hyödyntää koira-ihminen -sidoksen osiakin hämmentävän tehokkaasti osoittaessaan ihmiselle paikan koira-ihminen -suhteessa.

2.2.6 Geneettinen koira-ihminen –sidos

Ihmiset ovat osallistuneet aktiivisesti koirien eugeeniseen kampanjointiin, joka laventaa pohdintaamme geneettisen koira-ihminen -sidoksen osa-alueelle. Tietoisesti rotupuhtaat koiraryhmät käyttävät hyväkseen ihmistieteistä tuttua rotuoppia. Maallikot ovat osallistuneet aktiivisesti tähän toimintaan samoin menetelmin kuin hankkeissa, joissa on tähdätty ihmiskunnan geneettisiin isolaatioihin.

Suomessa ovat erityisesti uskonnolliset seudut Pohjanmaalla ja Sisä-Savossa olleet aktiivista aluetta tällä alalla. Rotuharrastajien asiassa tekemä työ on harvinainen esimerkki maallikoiden tekemästä tieteellisestä työstä, koira-ihminen -suhteen laventamisesta eliösuhteesta mentaalisuhteeksi. Tekijät haluavat julkaisulla muistuttaa rotujen historiallisesta merkityksestä sekä tieteen nykytilasta Suomen 100-vuotisjuhlallisuuksiin liittyen.

27

Esimerkiksi Suursuomenpystykorvan rotu perustettiin jo 1892. Aktiivinen siitostyö oli huipussaan 1920- ja 1930-luvuilla, kunnes hiipui viime sotien jälkeen. Perussuomenpystykorvan siittäjät ovat kuitenkin viime aikoina aktivoituneet, sillä tämän rodun pohjana ollut Suursuomenpystykorvan rotu oli lähellä kuolla sukupuuttoon.

Vuonna 1966 rotu herätettiin henkiin Ruotsissa kourallisesta koiria. Ruotsissa eugeniikan perusteet ovat vahvat. Tässäkin kansallisesti tärkeässä asiassa kansainvälinen yhteistyö on ollut tärkeää, tässä tapauksessa yhteistyö ruotsalaisten kanssa on ollut vilkasta. Perussuomalaisten rotujen säilyttäminen on siten pystytty pelastamaan viime vuosikymmenten aktiivisella jalostustyöllä. Ennakkoluulotonta asennetta tarvitaan jatkossakin.

Kansallisesti merkittävät koirarodut tulevat säilymään vain vaalimalla rodunomaisen käytön jatkamista. Tämä seikka yhdistää kaikkia potentiaalisia geneettisiä isolaatioita globaalissa ekosysteemissä. Nykyiset myyttiset kansalliset arvorodut ovat ikävästi sekoittumassa postmoderneihin, ehkä jopa alempiarvoisiin, rotuihin.

Suomalaisilla on sisäsiittoisessa jalostamisessa erityisvastuu. Ilman haukkuvia koiria näyttäisi itsenäinen Suomi kovin erilaiselta vuonna 2017 ja, uskallamme väittää, myös tulevina vuosina ja vuosikymmeninä.

3 Koira-ihminen –yhteisö

3.1 Sisätila

3.1.1 Koira-ihminen –suhteen koira sohvalla

Koiralle on kehittynyt merkittävä rooli koira-ihminen –suhteessa asuintilan haltuun ottajana, ja koira-ihminen –suhteen koira saattaa jopa tervehtiä asuntoon ja asuintilaan tulevaa vierasta ensimmäisenä, heti asuintilan tai asunnon ovella.

Vaikka tämä käytäntö, koira osoittamassa hallitsevansa tai suorastaan omistavansa asunnon tai asuintilan, on yleinen ilmiö, voi se myös olla hämmentävää sellaisen ihmisen näkökulmasta, jolla ei ole erityistä koira-ihminen –sidosta kyseiseen koiraan tai koiriin yleensä. Hämmennystä ei lueta tässä tapauksessa eduksi tai sivistyneeksi käytökseksi.

Koira on yllättävän monessa, alustavien selvitysten mukaan jopa enemmistössä koira-ihminen –suhdetta, vallannut paikkansa koira-ihminen –yhteisön keskeiseltä paikalta oleskelutilassa: sohvalta (Kuva 9).

Sisätiloissa koira-ihminen –sidos on harvoin käytössä, joten koira voi näissä tapauksissa soveltaa epäsuoraa valta-asemaansa, jonka voi tulkita olevan koira-ihminen –sidoksen jatkumo ja luovaa määräävän aseman käyttöä koira-ihmisen –suhteessa. Rajatussa tilassa voi olettaa, että koira säilyttää valta-asemansa, vaikka fyysinen koira-ihminen –sidos on katkaistu, kuten runsaasti tutkimusresursseja saanut sohvakäytöstutkimus selkeästi osoittaa.

Kuva 9. *Koira sohvalla, kohtuullisen tyypillinen tilanne koira-ihminen -suhteen koiralle. Koira ottaa melko usein vähintään 2x tilan sohvalta verrattuna koiran kokoon ja lajityypilliseen sohva-asentoon. Juuri julkaistun tutkimuksen mukaan koiran ottama tila koira-ihminen -suhteen ihmisen omistamalla sohvalla on ollut suurimmillaan 3.25x koiralle oletettu vähimmäisviihtyvyystila. Kuva CC BY 4.0.*

Sisätiloissa koira-ihminen -suhde määräytyy monien muidenkin kuin koira-ihminen -sidoksen rajaamien seikkojen ohjaamana, joten emme tässä yhteydessä puutu tähän erikoistapaukseen muutamaa seuraavana esitettyä esimerkkiä laajemmin. Koiralla on selkeä rooli sisätilojen napisijana, joka sekä ohjaa ihmisen käyttäytymistä ja antaa tarkkoja ohjeita sekä vinkkejä ympäristön tilasta.

Nämä sisätilojen koira-ihminen -suhteen käyttäytymismallit ovat, niin kuin jo aiemmin todettiin, johdettavissa koira-ihminen -sidoksesta ja koira-ihminen -suhteen kehittymisestä. Palaamme näihin kokonaisuuksiin myöhemmissä selvityksissä.

3.2 Ulkotila

3.2.1 Koira-ihminen –suhteen koira tervehtimässä vieraita ohikulkijoita

Usein havaitaan, että koira-ihminen -suhteen kokonaisuuksien kohdatessa koira ottaa aloitteen ja lähestyy kohtaamistilanteessa joko koiraa, ihmistä tai jopa koira-ihminen -suhteen kokonaisuutta ensin ja aktiivisesti. Tähän saattaa liittyä myös ääntelyä, jota on äänellis-toiminnallisesti luonnehdittu esim. räksytykseksi.

Tätä ääntelyä on usein selitelty sillä, että koira on useassa koira-ihminen -suhteen ulkoiluun liittyvässä sosiaalisessa tilanteessa omaksunut kohtaamistiedottajan roolin. Koira voi ilmaista sekä toisen koiran, ihmisen, että koira-ihminen -suhteen kokonaisuuden kohtaamisen mahdollisuuden koira-ihminen -suhteeseen liittyvän ulkoilun yhteydessä usealla eri tavalla, jotka poikkeavat selkeästi toisistaan ja ilmaisevat koira-ihminen -suhteen laatua selkeän eriävaisesti.

Koiran omaksuttua selkeän johtajan roolin koira-ihminen -suhteessa on tavanomaista, että koiran viestitys kontaktista toiseen koiraan, ihmiseen tai koira-ihminen -suhteen kokonaisuuteen on maltillista, mutta päättäväistä. Tähän liittyy koiran huomion kiinnittyminen, usein hienovaraisella tavalla, potentiaaliseen kontaktiryhmään.

Koiran viestitettyä tilanteen hallintaa ja ymmärrystä koira-ihminen -suhteen ihmiselle, se voi jopa osin suunnata huomionsa koiran kannalta oleellisiin, mutta kohtaamisen kannalta toissijaisiin asioihin ja toimintoihin. Koira on usein koira-ihminen -suhteen kohtaustilanteissa itse kohtaamistilanteen herra (Kuva 10).

Kuva 10. *Koira voi ottaa aktiivisen roolin tervehtiessään koira-ihminen -suhteen ihmisen tuttuja tai tuntemattomia lajikumppaneita ulkotilassa. Kuvassa koira-ihminen -suhteen koira on ottanut tervehtimistilanteen haltuun. Taustalla koira-ihminen -suhteen ihminen seuraa tilannetta rauhallisena, niin kuin tämänkaltaiseen jokapäiväiseen koira-ihminen -suhteen koiran omimaan tervehtimistilanteeseen tulee suhtautuakin. Kuva CC BY 4.0.*

Toisessa ääripäässä on paniikinomainen reaktio, jonka aikana koira hämmentää joko tarkoituksellisesti tai epäonnistuneen tilannehallinnan vuoksi koira-ihminen -suhteen ihmisen ja mahdollisesti myös kohdattavan koiran, ihmisen tai koira-ihminen -suhteen kokonaisuuden. Tällöin koira-ihminen -suhteeseen liittyvään kohtaamistilanteeseen liittyvään viestintään voi kuulua äänekästä ääntelyä ja joskus jopa aggressiota ilmaisevaa käyttäytymistä.

Tämä koira-ihminen -suhteen kohtaamistilanteeseen liittyvää ilmaisuun kuuluva, tilannetta painottava käytös on tulkittu usealla eri tavalla. Suurin osa koira-ihminen -suhteen kohtaamistilanteen tarkkailuun liittyvästä aineistosta viittaa siihen, että koira pyrkii tällä ylenmääräiseksikin luokiteltavalla käyttäytymismallilla hajottamaan sekä omaa koira-ihminen -suhdettaan että kohdattavaa koiraa, ihmistä tai koira-ihminen -kokonaisuutta. Hämmentämisen syistä on useita, osin ristiriitaisia teorioita.

Osa selvityksistä kertoo, että koira ottaa tämänkaltaisissa koira-ihminen -suhteen tilanteissa tilanteen haltuun ja pyrkii korostamaan rooliaan tilanteen hallitsijana. Tämän koulukunnan mukaan koira on käytöksessään johdonmukainen ja pyrkii vain vahvistamaan rooliaan tilanteen hallitsijana toiminnalla, jolla ei pikaisen tarkastelun mukaan ole järkevää perustetta.

Motiivi tämänkaltaiselle, pinnallisesti harhaiselle toiminnalle on koira-ihminen -suhteessa kohtaamistilanteen johdattaminen koiran kannalta haluttuun suuntaan muiden kohtaamistilanteessa mukana olevien kustannuksella. Toisen koulukunnan mukaan tämä käyttäytyminen koira-ihminen -suhteen kohtaamistilanteessa on harhauttavaa ja kertoo koiran heikosta roolista koira-ihminen -suhteessa, jolloin koiran kannalta voi olla tarkoituksenmukaista hämmentää koira-ihminen -suhteen ihmistä ja välttää tilanne, jossa ihminen voisi johdatella kohtaamistilanteen kehittymistä.

3.2.2 Koira-ihminen –suhteen koira aloitteen tekijänä tervehtimisessä

Koira-ihminen -suhteen yhteisössä on usein nykyisin tapana, että koira tervehtii yhteisöön saapuvaa ulkopuolista henkilöä ensimmäisenä. Tästä tervehtimisjärjestyksestä on kaksi periaatteellista muotoa, jotka perustuvat koiran sijoittumiseen koira-ihminen -yhteisön asuttamassa tilassa.

Muodoista aiemmin kehittyneessä koira on koira-ihminen -yhteisön asuttaman tilan siinä osassa, josta käytetään nimitystä 'ulkona'. Myöhemmin kehittyneessä toisessa variaatiossa koira on koira-ihminen -yhteisön asuttaman tilan siinä osassa, josta käytetään nimitystä 'sisällä'. Nämä nimitykset on johdettu suoraan konkretiasta ja kuvaavat ulkotiloja (ulkona) ja sisätiloja (sisällä). Seuraavassa käsitellään näitä kahta perustapausta erikseen, niiden oletetussa ajallisessa kehittymisjärjestyksessä.

Koiran ollessa ulkona sen tehtäviin on kuulunut koira-ihminen -suhteen koiran aiemman alisteisen aseman vuoksi vartiointi. Tätä tehtävää, niin kuin muitakin koiralle aiemmin koira-ihminen -suhteessa määrättyjä tehtäviä ja velvollisuuksia, on koira myöhemmin käyttänyt hyväkseen koira-ihminen -suhteeseen liittyvissä valtataistelussa. Tästä erikseen muualla.

Koira on siis tehtävänsä, vartioinnin, luonteen mukaisesti tervehtinyt koira-ihminen -yhteisöön tulijaa ennen yhteisön muita jäseniä. Tämä on toiminut varoituksena, eräänlaisena alkukantaisena ovikellon ja ovisilmän yhdistelmänä, ja on siten edistänyt koiran roolia koira-ihminen -suhteeseen liittyvän yhteisön valtarakenteessa.

Alkuvaiheessa koira ei ollut koira-ihminen -suhteeseen liittyvään yhteisöön tulijan kanssa ruumiillisessa kontaktissa, vaan tervehti koira-ihminen -suhteeseen liittyvään yhteisöön tulijaa matkan päästä (*varoetäisyys*). On esitetty, että koira-ihminen -suhteen koira oli tässä vaiheessa vielä niin aggressiivinen, että ruumiillinen kontakti ei ollut tarkoituksenmukaista.

Koira oli kuitenkin jo varhaisessa vaiheessa omaksunut koira-ihminen -suhteeseen liittyvän yhteisön tavaramerkin, jopa brändin, roolin ja on oletettu, että lähellä toisiaan sijaitsevat erilliset, koira-ihminen -suhteeseen liittyvät yhteisöt voitiin erottaa toisistaan koiran tervehdinnän perusteella. Tämä vähensi tarpeettomien vierailujen määrää ja auttoi tulijaa suunnistamaan myös vaikeissa oloissa kohti kyseiseen hetkeen liittyvää tai muun toiminnan kannalta oleellista koira-ihminen -suhteeseen liittyvää yhteisöä. Tämä on tietenkin edellyttänyt hyvää äänien erottelukykyä, joten tämä teoria koira-ihminen -suhteen koirasta historiallisena suunnistamisen apuna on herättänyt ajoittain vilkasta akateemista keskustelua.

Koira valtasi myöhemmin paikkansa koira-ihminen -suhteeseen liittyvässä yhteisössä keskeiseltä, sisätilaan liittyvältä kulttiesineeltä, sohvalta, jota merkittävää valtasuhteisiin liittyvää tapahtumaa on edeltänyt pitkä sohvakaappaukseen liittyvä valmistelu, josta hieman seuraavassa.

Tätä ennen koira-ihminen -suhteen koira oli osoittanut riittävää valppautta vartiointitehtäviin myös ollessaan sisätiloissa, ja tämä yhdessä koiran onnistuneen pitkäaikaisen manipulaation kanssa varmisti koiran paikan koira-ihminen -suhteeseen liittyvän yhteisön ihmispuolen ryhmässä. Tätä kutsutaan *koiran humanisaatioksi*.

Yhtenä merkittävänä seurauksena tästä merkittävästä saavutuksesta koira-ihminen -yhteisössä sisätiloissa oleileville koirille on langennut itsestään selvänä tehtävänä koira-ihminen -suhteeseen liittyvään yhteisöön tulijan tervehtiminen sekä asettamalla tervehtimiseen liittyvän *tervehtimisrituaaliääntelyn* (murina tai haukkuminen) tilanteen mukaan, että ottamalla ruumiillista kontaktia.

Koira-ihminen -suhteeseen liittyvään yhteisöön tulijat ovat usein omaksuneet koiran roolin vierailemassaan koira-ihminen -suhteeseen liittyvässä yhteisössä ja tervehtivät sekä sanallisesti että ruumiillisesti koira-ihminen -suhteen koiraa ennen koira-ihminen -suhteeseen liittyvän yhteisön ihmisiä.

On painotettava tässä yhteydessä, että ihmisyhteisöön tultaessa on perinteisesti ollut tapana tervehtiä sitä läsnäolijoista, joka on ihmisyhteisön (vrt. koira-ihminen -yhteisö) hierarkiassa korkeimmalla. Tästä on toki useita variaatioita ja joskus on vaikea tietää, kuka vastaanottavista ihmisistä on kunakin hetkenä tärkein. Tätä ongelmaa ei ole, jos koira ottaa rohkeasti johtajan roolin ja viivyttelemättä tervehtii koira-ihminen -suhteeseen liittyvään yhteisöön tulijaa.

Joissakin tilanteissa koira-ihminen -suhteeseen liittyvän yhteisön jäsenten tervehtimisjärjestys on epäselvä. On mahdollista, että koira-ihminen -suhteeseen liittyvään yhteisöön tulija ei ymmärrä koiran roolia koira-ihminen -suhteeseen liittyvässä yhteisössä ja tekee virheitä mm. väistelemällä koira-ihminen -suhteen koiraa ja pyrkimällä tervehtimään koira-ihminen -suhteeseen liittyvän yhteisön ihmisiä ennen koiraa.

Tämä epäjärjestystä aiheuttava käyttäytymismalli on luonnollista eläinten välisessä vuorovaikutuksessa, tosin arvokkaaksi tarkoitettu vierailu koira-ihminen -suhteeseen liittyvään yhteisöön voi saada heti alkuun irvokkaita piirteitä väärinkäsitysten vuoksi.

Koira-ihminen -suhteen koiran rooli tervehtijänä voi auttaa muita koira-ihminen -suhteeseen liittyvän yhteisön jäseniä selvittämään oman yhteisön tai tulijan suhdetta koira-ihminen -suhteen koiraan ja koira-ihminen -suhteeseen liittyvään yhteisöön. Vuorovaikutustapausten seurannassa on

kyetty erottamaan toisistaan muutama koira-ihminen –suhteeseen liittyvän yhteisön ymmärtämisen kannalta oleellinen ihmistyyppi.
On havaittu, että *koiraihminen* (käytetty nimitystä *I-ihmistyyppi*) ymmärtää koira-ihminen –suhteeseen liittyvän yhteisön monimutkaisen vuorovaikutusverkon vaivatta ja onnistuu vastaamaan koira-ihminen –suhteen koiran tervehdyksiin sekä sanallisesti että ruumiillisesti.
Ihmiset, jotka eivät määritelmällisesti kuulu *koiraihmisiin* (käytetty nimitystä *II-ihmistyyppi*), toimivat vastaavassa tilanteessa satunnaisen järjenvastaisesti ja voivat jopa ohittaa koira-ihminen –suhteen koiran tervehtimisen tullessaan koira-ihminen –suhteeseen liittyvään yhteisöön.
Samalla on huomautettava, että tässä kappaleessa määritettyyn I-ihmistyyppiin kuuluviin henkilöihin sisältyy myös *pseudo I-ihmistyypin* henkilöitä, jotka tässä tarkemmin määrittelemättömistä syistä vai luulevat kuuluvansa I-ihmistyyppiin ja siten, aiheuttamalla sekaannusta tulkintoineen, toimivat koirien tarkoitusperien hyväksi. Pseudo I-ihmistyyppiä on kuvattu myös koira-ihminen –suhteen myyriksi, joskus jopa vain paikallis- ja valtakunnanpolitiikan ulkopuolelle jätetyiksi hylkiöiksi.

3.2.3 Koira-ihminen –suhteen ihminen koiran jätösten muodostumisen seuraajana

Koira-ihminen –suhteen vuorovaikutuksessa on erittäin mielenkiintoinen, joidenkin käyttäytymistutkijoiden mietiskelynomaiseksi väittämä ulottuvuus, jota määrittelee mm. ihminen tuijottamassa keskittyneenä, jossain tapauksissa tupakoiden tai pieni muovipussi kädessä, koiran ulostamista.

37

Tätä mietiskelyhetkeä nimitetään länsimaista mietiskelykulttuuria käsittelevissä teoksissa, ei kuitenkaan kaikissa ja yksimielisesti, *I tyypin mietiskelyhetkeksi*. Tämä mietiskelyhetki voi kestää jopa minuutteja, riippuen koiran tarpeesta rauhoittaa tilanne ja antaa ihmiselle mahdollisuus koota ajatuksensa.

Tämänkaltainen yhden koira-ihminen -suhteen ihmisen aloittama mietiskely voi jopa johtaa joukkomietiskelyyn, jossa usea ihminen (koira-ihminen -suhteen ihmisiä tai yksittäisiä ihmisiä) keskittyy hiljaisena seuraamaan katseellaan - mikäli oletetaan, että tässä mietiskelytilassa ihminen voi tehdä näköhavaintoja - koiran ulostamista tai ulostetta.

Tämänkaltaisen mietiskelyhetken on tarkoissa seurantakokeissa todettu voivan keskeytyä niin kuin toisarvoiseen seikkaan, kuin vallitsevaa säätä koskevaan lausahdukseen. Usein kuitenkin nämä mietiskelyhetket päättyvät jätöstä koskevan havainnon ääneen lausumiseen ja, valistuneemman ihmisyksilön osalta, koiran jätöksen talteenottoon.

Koiran ulosteen talteenotto on herättänyt hämmästystä ja sen on oletettu liittyvän esimerkiksi ravintovarojen talteenottoon, energiavarojen kartuttamiseen tai jopa mietiskelyyn liittyvien muistoesineiden tallentamiseen ja mietiskelyyn liittyvien muistiinpanojen täydentämiseen. Viime aikoina koira-ihminen -suhteen ihmiset ovat myös heti tuoreeltaan tutkineet jätöksen luonnetta päällisin puolin, ilmeisesti selvittääkseen koiran tilanteesta seikkoja, joita ei yleensä ole haluttu käyttäytymistarkkailukokeissa erikseen eritellä.

Tästä pitkästä mietiskelyhetkestä on kehittynyt *II tyypin mietiskelyhetki*, nimitys ei tosin ole vielä vakiintunut, vaikka sitä on kirjallisuudessa käytetty erotukseksi I tyypin mietiskelyhetkestä. Tämä II tyypin mietiskelyhetki on selvästi nopeampi rytmiltään kuin I tyypin mietiskelyhetki, vaikka II tyypin mietiskelyhetkikin on kehittynyt koira-ihminen -suhteen koiran jätösten ympärille. Tässä II tyypin mietiskelyhetkessä ihminen seuraa koira-ihminen -suhteen koiran nestemäisen jätöksen eritystapahtumaa. Tähän II mietiskelyhetkeen voi myös liittyä tupakointi, mutta harvemmin muovipussiin tai muuhun keräimeen liittyvät toiminnot. II tyypin mietiskelyhetken jälkeiset toimet liittyvät usein nestemäisen jätöksen värimaailman pikaiseen tarkasteluun ja, tosin vain talvisin ja silloinkin vain hyvin harvoin, vaivihkaiseen nestemäisen jätöksen peittämiseen vasemman tai oikean jalan ulko- tai sisäsyrjällä siirrettävällä lumimassalla. On esitetty, että II tyypin mietiskelyhetki on kehittynyt näennäisen ajansäästön motivoimana, niin sanottuna pikamietiskelymenetelmänä, jonka esikuvana on energiavarojen täydentämiseen liittyvä voimauni (englanniksi 'power nap').

3.2.4 Ihmisten välisten keskustelujen aloitus koira-ihminen –suhteeseen liittyvässä ulkoilutilanteessa

Modernissa kaupungistuneessa ympäristössä perinteiset säätilaan ja satoennusteeseen perustuvat puheenaiheet ihmisten välisen vuorovaikutuksen ensiaskeleina ovat osin vanhentuneita ja jopa leimaavia. Koiralta tämä ei ole jäänyt huomaamatta ja koira-ihminen -suhteen koira onkin ottanut merkittävästi määräävämmän roolin ihmisten välisen vuorovaikutuksen aloitusrituaaleissa. Tämä koskee myös puheenaiheiden määritystä ja rajaamista.

Toisilleen aiemmin tuntemattomien ihmisten on, ainakin toisen osapuolen kiinteän koira-ihminen -suhteen ansiosta, kohtuullisen helppoa ja jopa yksinkertaista aloittaa sanallinen vuorovaikutus ottamalla toisen tai molempien kohtaavien ihmisten koira-ihminen -suhteen koira esille heti keskustelun aluksi.

Tämä lyhentää henkilökohtaisuuksien menevään keskusteluun vaadittavaa aikaa merkittävästi verrattuna sanallisen vuorovaikutuksen aloittamista jo tässä kappaleessa aiemmin mainituista aiheista, kuten säätilasta tai satoennusteesta. Koira-ihminen -suhteen koiran on usein näissä tilanteissa havaittu ottavan tarkkailijan roolin, alkuun vaihdettujen tervehdysten jälkeen, jolloin koira-ihminen -suhteen ihmiselle tulee ikään kuin automaattisesti lisää tilaa koira-ihminen -suhteessa. Tämä tosin vaatii sitä, että koira-ihminen -suhteen koira voi näissä vuorovaikutustilanteissa luottaa koira-ihminen -suhteen ihmiseen.

Koira-ihminen -suhteen koira on joissakin tapauksissa ottanut merkittävän roolin koira-ihminen -suhteen ihmisen parittamisessa (Kuva 11). Kirjallisuudessa kuvataan useita tapauksia, yksityiskohtaisestikin, joissa koira-ihminen -suhteen koira on merkittävästi edistänyt koira-ihminen -suhteen ihmisen mahdollisuuksia sekä pariutumiseen että parittelemiseen toisen ihmisen kanssa.

Näissä pariutumis- ja parittelutilanteita edistävissä alustavissa ihmisten välisissä vuorovaikutustilanteissa ovat koira-ihminen -suhteen ihmisen keskustelut koirastaan muiden ihmisten kanssa merkittävässä roolissa.

On voitu osoittaa, tosin laskelmia on myös arvosteltu, että tietyt myönteiset keskustelukokonaisuudet kahden ihmisen välillä toisen osapuolen koira-ihminen -suhteen koirasta ovat voineet edistää pariutumis- ja parittelumahdollisuuksia kaksin-, jopa kahdeksankertaisiksi, verrattuna kontrollina käytettyihin ihmisten välisiin keskusteluihin vastaavissa tilanteissa, mutta ilman koira-ihminen -suhteen koiran vaikutusta keskusteluihin.

Kuva 11. *Koira-ihminen -suhteen ihmisen parittelumahdollisuudet oman lajinsa kanssa voivat parantua koira-ihminen -suhteen koiran ottaman aktiivisen roolin ansiosta. Kuvassa koira-ihminen -suhteen ihminen on, todennäköisesti lajikumppaninsa kanssa pariteltuaan, tiineenä (nuoli). Tässä teoriassa on huomattavia aukkoja, sillä on todettu koira-ihminen -suhteen ihmisten tulleen tiineeksi myös ilman lajityypillistä parittelua, tästä on olemassa sekä kansantarujen että modernin lääketieteen esittämiä näyttöjä. Kuva CC BY 4.0.*

On jopa esitetty, joskin teoriaa on verisesti ja kiihkeästi arvostelu, että ihmiskunnan elinkelpoisuus on parantunut koira-ihminen -suhteen koiran edistämän pariutumis- ja paritteluvaikutuksen vuoksi ja tämä on osaltaan estänyt ihmisen sukupuuton, ainakin toistaiseksi. Vaikka koiran vaikutus koira-ihminen -suhteen ihmisen elossa säilyvyyteen ja geeniperimän siirtoon ei olisikaan sukupuuttoa estävä, on koira-ihminen -suhteen koiran roolin ihmisen parittajana kaikesta kritiikistä huolimatta ollut ilmeisen merkittävä.

Merkillepantavaa on myös se, että on koira-ihminen -suhteen koiran kannalta oleellista edistää niiden koira-ihminen -suhteen osapuolten (yleensä ihminen) hyvinvointia, jotka suhtautuvat myönteisesti koiran koira-ihminen -suhteen ihmiselle tarjoamiin pariutumismahdollisuuksiin.

Muuttuvamittainen koira-ihminen -sidos antoi koiralle myös mahdollisuuden uusien yllättävien tilanteiden aiheuttaman shokkiarvon hyödyntämiseen. Muuttuvamittaisen koira-ihminen -sidoksen ansiosta koira pääsi ohjailemaan koira-ihminen -suhteen ihmisen parittelukumppanin valintaa aiempaa suuremmasta reviiristä, jopa kaupungistuneen ympäristön kapeiden katujen vastakkaiselta puolelta.

Tämän laajentuneen reviirin ja ihmisen parittelumahdollisuuksien yhtäkkisen lisääntymisen vaikutusta koira-ihminen -suhteen muuntumiseen on laajasti, joskin tulosten osalta hämmentävästi, tutkittu ilman, että tähän liittyvillä teorioilla olisi laajempaa kannatusta. Rahoituksen vähäisyys ja väärä kohdentaminen ovat tämänkin tutkimusalueen harmeina.

Koira-ihminen -suhteen tärkeyttä edellä kuvatun kaltaisessa vuorovaikutuksessa on korostanut laaja ja pirstaleinen, osin myös tässä yllättävä koulukunta (usein käytetty nimitys terveysnatsit tai -fasistit), jonka mukaan ihmisten pariutumisrituaalien siirtyminen synkeistä sisätiloista (esim. baarit ja ravintolat), joissa ollaan tekemisissä harmillisten kemikaalien kanssa (esim. alkoholi, tupakka ja huumeet), ulkotiloihin, on ollut omiaan edistämään kansanterveyttä. Terveysfasistien alalahko, joka on suuntautunut kieltämään tupakoinnin, on tosin onnistunut tuhoamaan osan tästä muutoin pätevästä johtopäätöksestä.

4 Koira-ihminen –suhdemuutunnan harhaisuuksia ja niiden käytännön seurauksia

4.1 Koira-ihminen –suhteen koiran hämmentävä rooli sisätiloissa

Koira-ihminen -suhteen koira usein tervehtii vierailijaa ensimmäisenä ja osoittaa siten olevansa laumansa, koira-ihminen -suhteen perheyhteisön, johtaja. Vierailija myös usein tervehtii koira-ihminen -suhteen koiraa ensin, vaikka tuttavuus perheyhteisön ihmisten kanssa olisi ollut selvästi pitemmältä ajalta ja jopa syvällisempi. Tämä voi aiheuttaa sekaannusta sekä koira-ihminen -suhteen yhteisössä vierailevan ihmisen roolissa kyseisessä perheyhteisössä, että vierailijan näkemyksessä omasta persoonastansa.

Selvitysten mukaan on mahdollista, että vierailija voi tilanteen hämmentämänä joutua persoonallisuuskriisiin pohtiessaan koira-ihminen -suhteen yhteisön uroskoiran mahdollista roolia omana vanhempanaan verrattuna passiivisesti sohvalla makaavaan koira-ihminen -suhteen yhteisön ihmisurokseen, joka ei reagoi vierailijan läsnäoloon.

Tilannetta sekoittaa usein muiden koira-ihminen -suhteen ihmisten tarinat siitä, kuinka sohvalla makaava ihmisuros

43

on itse asiassa kyseisin ihmisyhteisön johtaja ja jopa vierailevan henkilön toinen vanhempi. Koira-ihminen –suhteen koiran on näissä tapauksissa usein todettu aktiivisesti puuttuvan tilanteen kulkuun, ja sotkevan kokonaiskuvan hahmottamista yrittävän vierailijan päättelyä mm. kiittämällä vierailijan huomion koiraan kokonaistilanteen asemesta. Tässä tapauksessa koira-ihminen –suhteen koira on onnistunut sotkemaan yksinkertaisen asetelman ihmisyhteisön sisäisestä asetelmasta ja tasapainosta ja jopa ottanut koiran roolin asemesta inhimillisen roolin.

4.2 Koira-ihminen –suhteen koira tekosyynä autokaupassa

Koira on, ulkopuolisen tarkastelun perusteella, omaksunut määräävän roolin useissa koira-ihminen –suhteeseen liittyvässä käytännön tilanteessa, joista käytämme tässä yhteydessä esimerkkinä koira-ihminen –suhteen ihmisten autonvalintaan liittyviä tekijöitä. Pohdintamme perustuu uutiseen (HS 4.3.2017 ja 27.10.2017), joka omalla vääjäämättömällä tavallaan kertoo koira-ihminen –suhteeseen liittyvistä valtasuhteista. Koiran voi väittää olevan määräävässä asemassa auton muotoilua ja valintaan koskevissa tilanteissa, jotka ovat toisaalta toisistaan erilliset toiminnot, mutta liittyvät kiinteästi yhteen ja samaan jatkumoon.

Ensin esittelemme hieman taustaa. Markkinatutkimuksen mukaan 98,7 % koira-ihminen –suhteen ihmisistä käyttää autoaan koiran kuljettamiseen. Tähän liittyen on väitetty tutkimuksiin perustuen (HS 27.10.2017), että koira on vaikuttanut auton valintaan yhdeksän kertaa kymmenestä (90 % valintatilanteista!) koira-ihminen –suhteeseen liittyneistä autokaupoista.

Tämä koira-ihminen -suhteen koiran osuus on selvästi suurempi kuin samaisen suhteen ihmisten yhteenlaskettu osuus (yleensä 1 - 3 ihmistä). Se on myös merkittävämpi vaikutukseltaan kuin auton väri tai käyttövoima. Tässä yhteydessä on esitetty myös väite, että koira-ihminen -suhteen koiran osuutta auton valintaan ei yleensä havaitse koira-ihminen -suhteen ihminen, jonka huomion auton valintatilanteessa vie valintastressi erityisesti, mikäli auton hinta, koko tai väri ovat auton valinnan piilovaikuttajia.

Japanilainen autonvalmistaja Nissan on kehitellyt auton, joka on räätälöity erityisesti koira-ihminen -suhteen koiria ajatellen. Koiralle suunnitellun auton tavaratilaan on rakennettu ruoka- ja juoma-annostelijat koira-ihminen -suhteen koiraa varten, mutta koira-ihminen -suhteen ihmisen ravinnontarvetta ja ravinnonkäytön edistämistä autossa ei ole huomioitu.

Myös autoon nousemista on helpotettu ulosvedettävällä luiskalla, jotta koira-ihminen -suhteen koiran ei tarvitse loikkia tai hypätä, koira-ihminen -suhteen ihminen taasen ahtautuu perinteisesti avautuvasta ovesta, usein koira-ihminen -suhteen ihmisen kannalta erittäin epäergonomisesti vääntäytyen. Koira-ihminen -suhteen koira voi tarkkailla koira-ihminen -suhteen ihmisen tekemisiä etupenkillä erityisen *koirakamerateknologian* avulla ja samalla huomioida koira-ihminen -suhteen koiran kannalta haitallisen käyttäytymisen. Tämä tarkkailu on usein vaivihkaista koira-ihminen -suhteen ihmisen keskittyessä auton ajamiseen ja liikenteen tarkkailuun. Myös jutteleminen onnistuu ääniyhteyden avulla.

4.3. Koira-ihminen –suhteen oikeustoimikoira

Koiralla on esitetty olevan pääsääntöinen vastuunsiirto-oikeus koira-ihminen -suhteen ihmiselle. Tästä seuraa, että koira-ihminen -suhteen koiran aiheuttama vahinko on lähtökohtaisesti korvausvastuullista ja koira-ihminen -suhteen ihminen on korvausvelvollinen. Järjestely on koira-ihminen -suhteen koiran kannalta erittäin edullinen, jopa suhteettoman edullinen, sillä täysi vastuu koira-ihminen -suhteen sähläyksistä lankeaa koira-ihminen -suhteen ihmiselle.

On väitetty, että tämä on yksi merkittävimmistä esimerkeistä osoittamassa koiran ylivoimaisuutta, suoranaista ylivertaisuutta, koira-ihminen -suhteessa ja kertoo erinomaisesti koira-ihminen -suhteen koiran määräävästä asemasta. Tämä on eräissä tapauksissa aiheuttanut holtittomuutta koira-ihminen -suhteen koiran käyttäytymisessä, mikä on lähtökohtainen korvausvelvollisuusasetelma huomioon ottaen oikeastaan vääjäämätöntä.

Voidaan myös puhua koira-ihminen -suhteen koirasta vaaranlähteenä (oikeudellisesti 'vaaranlähdeajatus'), jolloin vastuu koira-ihminen -suhteen varsinaisilla osapuolilla voi olla jopa järjestyslaissa esitettyä vastuuta laajempi.

Hallinto-oikeudessa on käsitelty tapausta, jossa HO:n ratkaisu koski koira-ihminen -suhteen koiran pitämistä kotona sisällä ja sisään astuvan vieraan loukkaantumista sekä erityisesti sitä, millä tavalla koira-ihminen -suhteessa koiran omistajan tulee varmistua siitä, että vieras tietää koira-ihminen -suhteen koirasta ja sen aiheuttamasta vaarasta.

Koira-ihminen -suhteen koiriin liittyvä lainsäädäntö on hajautettu useaan eri lakiin. Lain edessä koira-ihminen -suhteen koira on esine, joka on yksi merkillinen piirre koira-ihminen -suhteen koiran aseman mieltämisessä. Koira-ihminen -suhteen koira voidaan siten ostaa ja myydä, ja tämä liiketoimi voidaan ainakin näennäisen pintapuolisesti tehdä koira-ihminen -suhteen ihmisen lähtökohdista. Tässä yhteydessä emme katso tarpeelliseksi puuttua koira-ihminen -suhteen koiran kykyyn ja tarpeeseen ohjailla koira-ihminen -suhteen ihmisen taloudellisia päämääriä erityisesti koira-ihminen -suhteen koiraa koskien. Kuitenkin on tärkeä huomata, että koira-ihminen -suhteen koiran omistaminen eroaa koira-ihminen -suhteen koiran kasvattamisesta.

Koira-ihminen -suhteen koiran omistaja on tyypillisesti yksityinen henkilö, kun koira-ihminen -suhteessa koiran kasvattaja on puolestaan yleensä elinkeinonharjoittaja (erityisesti koira-ihminen -suhteen koirilla liiketoimintaa harjoittava yrittäjä).

On myös huomattava, että virkakoirien erioikeudet hämmentävät ihmisiä, muitakin kuin koira-ihminen -suhteen ihmisiä. Määritelmän mukaisesti virkakoirat ovat koira-ihminen -suhteen yksi erityisryhmä, jolla on asema yhteiskunnan valtarakenteessa ja jopa koira-ihminen -suhteen ihmiseen verrattava virka tai toimi yhteiskunnan eri organisaatioissa. Koira-ihminen -suhteen virkakoirilla on oikeus ja mahdollisuus pakkokeinoihin ja väkivaltaan, tehostettuun valvontaan sekä ne osallistuvat muuhunkin toimintaan, joka katsotaan kuuluvan koira-ihminen -suhteen yleistä hyötyä edistäville virkakoirille.

Virkakoiraa ei kuitenkaan sido virkalain rajoitukset tehtävien hoitamisesta tuloksellisesti ja tarkoituksenmukaisesti. Tässä korostuu koira-ihminen -suhteen koirien erityinen asema, sillä samaista virkalakia ei myöskään sovelleta eduskunnan oikeusasiamieheen ja apulaisoikeusasiamiehiin

47

eikä eduskunnan kanslian, tasavallan presidentin kanslian, eduskunnan oikeusasiamiehen kanslian ja valtiontalouden tarkastusviraston virkamiehiin eikä myöskään Suomen Pankin ja Kansaneläkelaitoksen virkamiehiin ja toimihenkilöihin, ellei laissa toisin säädetä.

Mainittakoon vielä, että termiä 'paskalakki' (tai sen johdannaisia) ei voi ilman sopivaa tai erityistä rangaistusta käyttää virkakoirasta, riippumatta virkakoiraan liittyvästä koira-ihminen -suhteesta.

4.4. Koira-ihminen –suhteen koira on ihmistä viisaampi

Ensimmäiseksi on huomattava, pienenä joskin merkittävänä yksityiskohtana, että koira-ihminen -suhteen koira vieraili avaruudessa ennen ihmistä (Kuva 12). Tämän syvemmin ei tarvinne tarkastella teemaa, jossa käsitellään ja perustellaan väitettä: *koira on ihmistä viisaampi*, joka on yleistys koira-ihminen -suhteen koiran mahdollisesta älyllisestä potentiaalista eri tilanteissa, tietenkin koira-ihminen -suhteen ihmiseen verrattaessa.

Asiasta kiinnostunutta lukijaa haluamme kuitenkin viihdyttää tässäkin yhteydessä koko rahalla, vaikka tutkielmamme onkin maksuton. Tätäkin esittämäämme ja hyväksymäämme käytännön ristiriitaa (väittämä 'ilmaiseksi voi saada jotain hyödyllistä' on XXIV -tyypin harhaksi oletetun tilanteen mukainen harha, kun sovelletaan taloustutkimuksen yleisesti hyväksyttyä, joskin myös kiistanalaista, 'yleisharhaperiaatetta') voi myös käyttää todistamaan koiran ylivertaista asemaa koira-ihminen -suhteessa.

Kuva 12. *Neuvostoliittolaiskoira edelläkävijänä avaruuden valloittamisessa. Milloin ihminen on käynyt ensimmäisenä yleensä missään? Tämä klassinen esimerkki osoittaa vääjäämättömästi koiran ylivertaisen älykkyyden koira-ihminen -suhteessa. Koira menee, niin halutessaan, raketilla Kuuhun, ihminen yrittää saada potkukelkkaa etenemään sohjossa! Kuva CC BY 4.0.*

Koira-ihminen -suhteen koiran älykkyyttä ja sen seurauksena ihmisen toiminnan ohjaamista koira-ihminen -suhteessa on tuotu esille useassa tieteen tulosten kansanomaistamiseen perustuvassa teoksessa, tästä yhtenä merkittävänä esimerkkinä eurooppalaisen lehtihenkilön, Tintin, elämää ja työskentelytapoja käsittelevä tietosarjakuvakirjasarja. Tässä kyseisessä, laaja-alaisessa teossarjassa tuodaan käytännön esimerkkien avulla esille koira-ihminen -suhteen koiran (Milou) määräävä rooli ihmisen käyttäytymisen ohjaajana, jossa tulevat esille selkeästi esimerkiksi koiran toimintojen yhteinen seuranta, pariutumiskäyttäytymisen oh-

49

jailu sekä koiran turhautuminen ihmisen tyhmyyteen. Samoja aihekokonaisuuksia on tuotu esille myös muissa vastaavissa tieteen kansanomaistuksissa, joista mainittakoon esimerkkeinä seuraavat tietosarjakuvasarjat: Mustanaamio (Devil), Lucky Luke (RanTanPlan), Asterix (Idefix) ja Tenavat (Ressu).

Koiran on havaittu toimivan tunnustelijana useassa uusille maantieteellisille alueille tai naapurustoon suuntautuvassa koira-ihminen -suhteen seikkailutilanteessa. Tämä koira-ihminen -suhteen koiran ottama käyttäytymismalli, jonka voisi kuvata esimerkiksi tilanteen haltuun ottamiseksi, on viite koiran älykkyydestä.

On myös spekuloitu, joskin harvoin, että ihminen käyttää koira-ihminen -suhteen koiraa hyväkseen näissä tilanteissa, mutta tätä hyväksikäyttömallia on voimakkaasti kritisoitu sen perusolettaman, ihmisen älykkyysolettaman, hataruuden vuoksi.

Koira-ihminen -suhteen koira peittää jätöksensä, noin yleensä, vaikka se ei olekaan kovin tehokasta. On tosin toisaalla esitetty, että koira ei välttämättä tarvitse tähän toimintoon koira-ihminen -suhdetta, joten toiminto ei liene koira-ihminen -suhteen evolutiivinen sivujuonne. Niin tai näin, koiran tarkoituksena voi myös olla, peittämisen asemesta, jätöstensä tehokas levittäminen, jota auttaa maan kuopiminen jätösten välittömässä läheisyydessä.

Mitä kauemmaksi edetään tiiviistä kaupunkiympäristöstä ja mitä pohjoisemmaksi edetään maantieteellisesti, sitä vähemmän koiran jätösten noukkiminen, pussittaminen ja roskiin toimittaminen kiinnosta muutoin koirasta kiinnostunutta koira-ihminen -suhteen ihmistä. Kiinnostuksen asteella ei näyttäisi tässä tapauksessa olevan mitään tekemistä koira-ihmisen -sidoksen tyypillä (määrämittainen vai muuttuva koira-ihminen -sidos). Tätäkin on käytetty esimerkkinä koiran suuremmasta älykkyydestä ihmiseen verrattuna.

Kuva 13. *Klassinen, runsaasti tarkasteltu valokuva esimerkkinä koira-ihminen -suhteessa ilmenevistä ilmeisistä hallintasuhteista. Kuvan perusteella on voitu päätellä (vähintään 23 julkaistua tulosta, kaikki eivät tosin päädy edes samansuuntaiseksi tulkittavissa olevaan loppupäätelmään), että koiran parinvalinta on tarkoituksellista ja liittyy läheisesti koira-ihminen -suhteen koirapainotteisen hierarkiaosuuden muodostumisen helpottamiseen. Kuvassa tulee hyvin esille myös, eräänlaisena sivuviitteenä, mm. koira-ihminen -sidoksen pituuden vaihtelu ja koira-ihminen -suhteen ihmisen ote sidoksesta. Kuva CC BY 4.0.*

On myös esitetty, että koira on ymmärtänyt kaupungistumisen aiheuttaman tarpeen siisteydelle ihmistä paremmin,

vaikka koiran käytös jätösten peittämisessä lienee perua ajalta, jolloin kaupungistuminen ei vielä ollut alkanut. On myös esitetty, että koira on voinut ennakoida kaupungistumisen ja pyrkinyt esimerkillään vaikuttamaan koira-ihminen -sidoksen kautta koiraan kytkeytyneeseen ihmiseen.
Vielä, melkein lopuksi, yksi merkittävä yksityiskohta, suoranainen koira-ihminen -sidoksen ja -suhteen klassikko. Koira-ihminen -suhteen koiran ylivertainen älykkyys näkyy myös selvästi toisen maailmansodan aikaisessa valokuvassa (Kuva 13), jossa tunnistamaton, ilmeisen saksalainen nainen ja hänen seurassaan oleva, niin ikään ilmeisen saksalainen sotilas, ovat tiukasti kiinni koira-ihminen -sidoksella koiraan.
Valokuvaa on laajasti analysoitu ja päädytty hämmästyttävän yhdenmukaisiin päätelmiin näkökulmasta ja tutkimusasetelmasta riippumatta: koirat ovat kuin ovatkin tässä koira-ihminen -nelisuhteessa määräävässä roolissa. Viimeaikaiset selvitykset ovatkin keskittyneet kyseisen valokuvan koirien todelliseen rooliin ihmis-ihmis -suhteen muutoksissa tuona ajankohtana. Ensimmäiset tulokset ovat olleet hämmentäviä, joskaan eivät ole mitenkään vähentäneet kuvan henkilöiden vastuuta teoistaan.
Koira-ihminen -suhteen koiran älykkyys saman suhteen ihmiseen verrattuna on merkittävästi hukannut resursseja. Jokakeväinen tiedotusvälineiden yleisönosastot täyttävä koira-ihminen -suhteen koirien jätöksiä koskeva mielipidemyrsky on sanomalehtien yleisönosastokirjoitusten historian aikana tuhonnut suomalaisia metsiä vähintään 0.001 % vuotuisesta kokonaiskäytöstä määritettynä.
Puunkulutus on laskettu kasvavaa pystykuutiota kohtia, jolloin on otettu huomioon sanomalehtien yleisönosastopals-

toihin käytetty sanomalehtipaperi, yleisönosastoviestiaihioiden työstö ja puhtaaksi kirjoitus sekä kirjekuoret ja postimerkit. Nykyisin, paperin mediamerkityksen vähetessä, internetin kapasiteettia tuhlataan erään arvion mukaan noin 3 - 5 kertaa verrattuna koira-ihminen -suhteen koiran jätösten tunnepitoista käsittelyä edeltävään aikaan, mielipidekirjoituksiin käytetyn kirjainmäärän mukaan määritettynä. On myös esitetty arvioita tuhlatusta ajattelukapasiteetista näitä koira-ihminen -suhteen vivahteita käsitteleviä mielipidekirjoituksia laadittaessa, mutta tutkijat eivät ole päässeet yksimielisyyteen siitä, olisiko tätä ajattelukapasiteettia käytetty tuloksekkaammin, mikäli koira-ihminen -suhteen ihmiset olisivat keskittyneet yhteiskunnan kannalta tärkeämpiin aiheisiin.

Koiranjätöksiin keskittyvien koira-ihminen -suhteen ihmisten (toisin sanoan mielipidekirjoittajien) ajattelukapasiteetti on ollut tässä tapauksessa aivan oikeassa ja heidän ajattelukapasiteettinsa mukaisessa käytössä ja heidän ajattelutyönsä aivoille tuottama toiminto on ollut yksilötason hyötyä mietittäessä kannattavaa.

53

Communications in Expanded Facts of Universal & Calculated Knowledge (eFUCK):

Vol. I. Saarti & Kananen & Jussila. Moderni kaninis-humaani suhde: koira-ihminen -suhteeseen liittyvän sidoksen muutosvaikutus.

Vol. II. Jussila & al. Tumputa (Rapuooppera -sinkula). Saatavissa: www.youtube.com/watch?v=cFoMOC1ojmY.

www.ingramcontent.com/pod-product-compliance
Lightning Source LLC
Chambersburg PA
CBHW030509220526
45464CB00006B/2723

www.ingramcontent.com/pod-product-compliance
Lightning Source LLC
Chambersburg PA
CBHW020634220526
45464CB00001B/146

KALA RAMESH writes and teaches haiku, tanka, haibun and renku to children, undergrads and senior citizens. She created the 'Rasika' form, an eight-verse renku fashioned after Matsuo Basho's non-thematic style; and she is passionate about getting haiku painted on city walls.

SANJUKTAA ASOPA has had her haiku published in several well-known journals that have fetched her a few awards along the way. She believes in the journey and not in the milestones.

SHLOKA SHANKAR is a freelance writer, poet, and visual artist from Bangalore, India. She is the founding editor of Sonic Boom.

www.vpindia.co.in

Tzetzka Ilieva, USA: "Skype" TI 11.3; "did I tell you" WHR, December 2012; "learning to accept" *Gazing at Flowers*, HSASRA 2013. **p. 195.**

Vadivelrajan S.B, India: "waterfall" Katha eBook 2013. **p. 197.**

Vandana Parashar, India: "chilly morning" Triveni MI, January 2016; "arranged marriage" Triveni MI, January 2016. **p. 197.**

Vidur Jyoti, India: "shared silence" *Carvings on Dew*, PA 2015; "shrivelled leaves" *Carvings on Dew*, PA 2015; "another blast" *Carvings on Dew*, PA 2015. **p. 196.**

Vidya S. Venkatramani, India: "muddy puddle" AHG 4.3; "hushed dawn" CA, September 2015. **p. 199.**

Vinay Leo R., India: "salted caramel", CA, September 2014; "autumn dusk" MH 45.2; "four-leaf clover" PJ March 2014; "ink puddle" SMK November 2013. **p. 198.**

Vishnu P. Kapoor, India: "winter night" IHA Spring 2008. **p. 199.**

Vividha Bhasin, India: "war cry" CA YC, May 2014. **p. 216.**

William J. Higginson, USA: "the clock" Years Collected Haiku, Vol. 1, HP 1987; "light hearted" 16 Haiku from The Four Seasons in Santa Fe; "a gun in the wildflowers" TI, August 2001; "deep in the arroyo" MH 33.2; "one maple leaf" THN VI:8; "late autumn" Upstate Dim Sum 2006; **p. 200, 201.**

Yajushi (Vinodh Marella), India: "walking to Ganga" SH v4n1. **p. 201.**

Yesha Shah, India: "sea mist" IK 4; "third trimester" THN XVI.2; "election fever" THN XVI.3. **p. 202.**

Yu Chang, USA: "writing cursive" THN XV.2; "blinding snow" *Seeds*, RMP 2009; "back at camp" THN III.2; "just long enough" THN VII.10. **p. 203.**

Permission received from all the poets for the haiku reprinted here. Our deepest gratitude to the family members of John Edmund Carley, Laryalee (Lary) Fraser, Martin Lucas, Peggy Willis Lyles, William Higginson and Svetlana Marisova for giving us the permission. Our thanks also to Nick Virgilio Haiku Association for lending Nick's haiku.

Stanford M. Forrester, USA: "godless month" PR; "when a scarecrow" HCN; "day moon" KO 27.4; "moon viewing party" AL 2004. **p. 183.**

Suhit Kelkar, India: "mossy wall" THN XVII.4. **p. 184.**

Surbhi Grover, India: "lonely road" CA, January 2014. **p. 187.**

Susan Antolin, USA: "night sky" MH 46.2; "assigning" FR 37.1; "summer clouds" MH 45.1; "curriculum vitae" *Close to the Wind,* HNACA 2013. **p. 185.**

Susan Burch, USA: "half-empty cup" HSA's GBSC 2014; "in the belly" BO 4; "looking for love" SB 1; "crowded sidewalk" SB 2. **p. 188.**

Susan Constable, USA: "how easily" AC 33; "Mother's Day" MH 42.3; "knife cold" MH 44.1; "cloudburst" THN XII.1; "amber light" THN X.1; "winter branches" WL 7. **p. 186, 187.**

Susan Diridoni, USA: "step back" IUH WH 19. **p. 189.**

Susumu Takiguchi, UK: p. 189.

Svetlana Marisova, New Zealand: "winter breeze" *Language and Silence - selected poems of Svetlana Marisova,* KP 2014; "crashing waves" *Language and Silence - selected poems of Svetlana Marisova,* KP 2014; "summer surf" *Language and Silence - selected poems of Svetlana Marisova,* KP 2014; "bitter spring" NG 2.4. **p. 190.**

Tanvi Nishchal, India: "broken ladder" CA YC, May 2014. **p. 216**

Tanvi Shah, India: "crowded city" CA YC, January 2016. **p. 216**

Terri L. French, USA: "evening meditation" HO 12.2; "not getting the whole story" DH Cycle 10; "first snow" SB 3; "lingering fog" SB 3. **p. 191.**

Thiagarajan A., India: "road puddles" THN VIII.4; "autumn evening" THN VIII.2; "cold evening" THN IX.2; "new actors" MH 39.2; "house hunt" THN XI.2. **p. 192, 193.**

Tito, Japan: "the mountain lake today" BS 3.2. **p. 193.**

Tom Clausen, USA: "gone south" THN XVIII.1; "before sleep" RN v1n2; "reading her letter" MH; "rushing" BSP, May 1998. **p. 194.**

Tyrone McDonald, USA: "tunnel" MH 42.2. **p. 195.**

"all the times" DH Cycle 16; "mountain pass" INK 15; "munching cookies" FR 37.2. **p. 170–172.**

Sasa Vazic, Serbia: "growing silence" SH v10.n2. **p. 170.**

Scott Metz, USA: "only american" MH 40.1. **p. 173.**

Seshu Chamarty, India: p.173.

Shabbir Shaikh, India: p. 173.

Sheila Windsor, UK: "almost winter" HN October 2012; "if I spin" RN 2002; "red balloon" HBA 2001; "night train" RTB 2003. **p. 174.**

Shernaz Wadia, India: 'slow day" MI 37, "on wet earth" MI 37. **p. 175.**

Shloka Shankar, India: "summer heat" UtB 2014; "eraser" CKK 53; "hyphenated" DH Cycle 18; "rejection mail" UtB 2015; "typecast" BO 8; "chry" MO 4; "before and after" TZS 2015; "the next sneeze" UtB 2016. **p. 176, 177.**

Shobhana Kumar, India: "blue moon" HT October 2014. **p. 189.**

Shrikaanth Krishnamurthy, UK: "London train" CA, September 2014; "swirling leaves" *galaxy of dust,* TRMAELH 2015; "navaraatri" MI, January 2016; "left right left" MI, January 2016; "one by one" THN XXVII.1; "light drizzle" MH 46.1. **p. 178, 179.**

Singh R.K., India: "looking for shade" MA 3. **p. 175.**

Smayan Mohanty, India: "pouring rain" MLDH, August 2015. **p. 214.**

Sneha Mojumdar, India: "autumn morning" CA YC, January 2015, "sunny morning" CA YC, January 2015. **p. 215.**

Snehith Kumbla, India: "chained dog" WHR, January 2011; "camphor flame" THN XIII.2; "winter grass" WHR, December 2012. **p. 180.**

Sonam Chhoki, Bhutan: "clinic waiting room" THN XIII.1; "night border crossing" SHR 26; "stolen wombs" HN, January 2013. **p. 181.**

Sondra J. Byrnes, USA: "hearing it pull" FR 38.1; "zazen" TI 13.2; "sighing" DH Cycle 20. **p. 182.**

Sparsh Agrawal, India: "sunrise ..." CA YC, May 2016. **p. 215.**

sprite (Claire Chatelet), UK: p. 187.

Sreelatha Nair, India: "from a green mound" CA, September 2015; "thunderstorm" MI, January 2016. **p. 184.**

2007; "rustling leaves" THN XV.3; "blue jean patches" MA 30; "empty park" AC 33. **p. 158, 159**

Roberta Beary, USA: "autumn moon" PIHC 2012; "new year sake" SMK, January 2014; "all day long" *Pocket Change*, RMP 2000; "needle juniper" MH 43.3. **p. 160.**

Rochelle Potkar, India: "periwinkle sky" AHG 4.3; "traffic snarl" TSC, August 2015; "train window" TSC, August 2015. **p. 156.**

Rohan Kevin Broach, India: "five rupees" CA YC, January 2015. **p. 214.**

Rohini Gupta, India: "vacation's end" BR 19, August 2008; "across the page" WHR, August 2011; "mid sentence" *whiskers and purrs, a book of cat haiku,* 2016. **p. 164**

Ron C. Moss, Australia: "old horses" THN XVI.4; "starry night" SMK, December 2006; "dry billabong" WH, Marc6h 2015. **p. 161.**

Rosa Clement, Brazil: "orange harvest" HR, June 2011. **p. 164.**

Ruth Yarrow, USA: "food bank line" FR 29.2; "planting peas" MH 42.2. **p. 162.**

Safiyyah Patel, UK: "ruby studs" Atoms of Haiku, AU 2015. **p. 167.**

Samar Ghose, Australia: "the flutter" UtB 2015; "city park bench" SB 1; "the fishermen's song" SB 1; "waterfront bar" PJ 15; "salt free diet" SB 4. **p. 166, 167.**

Sandi Pray, USA: "quiet moments" AHG 3.1; "waterfall" UtB 2013; "one coyote" MLDH June 2015. **p. 168, 169.**

Sandip Chauhan, USA: "moonbow" IMBA 2014; "a torn page" MDN 2012. **p. 169.**

Sandra Martyres, India: p. 165.

Sandra Simpson, New Zealand: "spattering rain" TRSMHA, 2011; "the moon's apostrophe" HPA (UK) December 2013; "photos of her father" NG 1.1. **p. 165**

Sanjuktaa Asopa, India: "this puddle" SMK November 2013; "fingernail moon" SMK January 2014; "shanty town" THN XIII.3; "after all the things" THN XIV.2; "thrush song" WHR June 2015; "galaxy" FR 37.3;

Poornima Laxmeshwar, India: "orange sunset" UtB 2014; "night deepens" SB 1. **p. 149.**

Pravat Kumar Padhy, India: "temple bell" FR 36.2; "early morning" PR 49. **p. 149.**

Pruthvi Shrikaanth, UK: "robins chirp" CA YC, January 2015; "bombers moon" CA YC, May 2014. **p. 213.**

Purushotham Ravela Rao, USA: p. 147.

Quamrul Hassan, Bangladesh: "cricket fever" HHA: Hyaku Haiku 2016; "north wind" *Spring moon* 2011. **p. 150.**

Raamesh Gowri Raghavan, India: "silent night" SB 2; "puja flowers" SB 2; "autumn shadow" CR 29. **p. 151.**

Rajeshwari Srinivasan, India: p. 150.

Ramesh K., India: "starlit sky" THN VI.5; "rustle of palm leaves" THN XIV.2; "clouds drift" *pebble to pebble: a collection of haiku and tanka;* "carnival over" THN XIV.3; "sound of rain" THN XIII.3; "scent of jasmine" MH 40.3; "a heron's pulsing throat" AHG 4.2; "village fair" THN XIV.1; "back from my hometown" PR. **p. 152–154.**

Randy Brooks, USA: "cracked glass" FR 26.2; "October light" THN X.1. **p.157.**

Rebecca Lilly, USA: "evening rain" *Shadwell Hills,* BBP 2002; "fading day moon" MH 35.3; "the gravedigger staring" *Elements of a Life,* RMP 2014. **p. 155.**

Renée Owen, USA: "dusk" MAP 32; "evening chill" SP CC 2008. **p. 158.**

Richard Gilbert, Japan: "as an and you" RR 11.2; "dedicated to the moon" NO 1. **p.157.**

Rishabh Jain, UK: "trekking" CA YC, September 2014. **p. 213.**

Ritaj K, India: "people everywhere" CA YC, May 2014. **p. 214.**

Robert D. Wilson, Philippines: "memories" *Vietnam Ruminations* 2003; "endless summer" *A Lousy Mirror* 2012; "humid night" *Jack Fruit Moon,* METP 2009; "yesterday" *A Lousy Mirror* 2012. **p. 162. 163.**

Robert Epstein, USA: "star jasmine" THN XVI.3; "zen garden" HSA MA

Olivier Schopfer, Switzerland: "far from home" IUH WH 198, January 2015; "end of story" TI 15.1; "round the corner" BO 4 2014. **p. 135.**

Padma Thampatty, USA: "fallen leaves" FR 38.3. **p. 138.**

Pamela A. Babusci, USA: "changing kimonos" RHBA 2002; "i traveled" DOL #4; "no strength to pray" WL 9. **p. 136.**

Paresh Tiwari, India: "music box" DH Cycle 17; "raked leaves" FS 3; "chardonnay" TI 14.1; "thunderclap" SPHC, 2015; "fading daylight" MH 45.3. **p. 138, 139.**

Patricia Prime, New Zealand: "flea market" BS 24.2; "riverside" BS 20.2; "in the margins" BS 22.3. **p. 137.**

paul m., USA: "daffodil shoots" THN, VII.3; "a line borrowed" THN XII.2; "bulbs dividing underground" MAP 17; "rehearsing a lie" AAAAA 8; "with a cheap beach towel" FR 32.3; "spring light" MAP 10; "deep winter" MAP 9. **p. 140, 141.**

Paul W. MacNeil, USA: "paddle at rest" MH 30.3; "stepping stone" Montage: The Book, 7; "windless heat" SPHC 2001. **p. 145.**

Payal A. Agarwal, India: "between us" AHG 4.1. **p. 145.**

Peggy Willis Lyles, USA: "in spite" SPHC 2008; "lunch at the zoo" MH 32.1; "boarding call" *To Hear the Rain: Selected Haiku of Peggy Lyles,* Clothbound 2002; "into the night" *To Hear the Rain: Selected Haiku of Peggy Lyles,* Clothbound 2002; "country church" *To Hear the Rain: Selected Haiku of Peggy Lyles,* Clothbound 2002; "morning twilight" AC 10; "I shake the vase" *Thirty-Six Tones,* SPR 1999; "piano lesson" HPNC 1999 Haiku Senryu Tanka Rengay Contest; "Indian summer" THN III.10; "summer night" CI 4.4; "before" BS 2.2. **p. 142–144.**

Penny Harter, USA: "do I gather peonies" *A Spray of Dogwood,* HP 2007; "migrating butterflies" FR 16.1. **p. 146.**

Peter Newton, USA: "standing" AC 26; "yard sale" KASIHC 2011; "over" TI 10.2. **p. 147.**

Peter Yovu, USA: "a falcon dives" MH 41.1. **p. 148.**

Polona Oblak, Slovenia: "autumn wind" MH 43.2; "bark" THN XVI.1; "what i said" MH 44.1. **p. 148.**

Martin Lucas, UK: "the thyme-scented" PR 39; "summer river" PR 39; "morning sun" BS 13.3; "crescent moon" BS 13.3. **p. 124.**

Max Verhart, Netherlands: "falling apple" SMK September 2006; "autumn wind" SMK November 2005; "quiet Sunday" MH 34.1; "tin soldiers" FR 2000/1; "out of the haze" SMK March 2006. **p. 122, 123.**

Maya Lyubenova, Bulgaria: "herring clouds" 17th KKHC; "broken wall" NG 1; "the b-flat" UtB 2014;" gamma therapy ward" WHR January 2015. **p 125.**

Melissa Allen, USA: "spring rain" MH 42.3; "nothing" NG 3.2; "particles" R'r 11.2; "the smallest bone" AC 28; "autumn sky" FR 46.1. **p. 127.**

Michael Dylan Welch, USA: "dwindling fire" THN VI.9; "the waiter interrupts" *Shiki Haikusphere*, Japan, OPC 2007; "grocery shopping" *Fig Newtons: senryu to go*, PH; "warm winter evening" MDNC 2003; "crackling beach fire" THN VI.11. **p. 128, 129.**

Michael McClintock, USA: "having no thought" HGHHA 2006; "done for the day" THN VIII.3; "a broken window" HM 4.4; "a poppy" HM 5.1; "the moon" BDA 2002; "spring dream" THN XVII.3. **p. 130, 131.**

Michael Rehling, USA: "as if i cared" AHG 3.4; "pretending" Haigaonline com; "prayer beads"details unknown. **p. 128.**

Mihir Oak, India: "still mirror" CA YC, January 2015. **p. 212.**

Minal Sarosh, India: "rippling laughter" UN April 2012. **p. 126.**

Mykel Board, USA: "fiftieth birthday" FR 1997. **p. 126.**

Nick Virgilio, USA: "after the bell" MH 16.3; "autumn twilight" HW 1.1; "lily" AH 1.2; "Easter morning" AH 4.1; "taking a hard look" BMP 1988; "on the manuscript" TLP collection, *A Life in Haiku* 2012. **p. 132–134.**

Nikita Engineer, India: "winter night" CA YC, May 2014. **p. 212.**

Niranjan Navalgund, India: "key chain" SB 2. **p. 135.**

Owen Bullock, Australia: "my shadow" PR 18. **p. 136.**

Kumarendra Mallick, India: "autumn morning" WHR August 2009; "day of festivity" CH 16 2014. **p. 109.**

Kushal Poddar, India: p. 109.

Lakshmi Ramaswami, India: "colours of Holi" MLDH, August 2015. **p. 212.**

Laryalee (Lary) Fraser, British Columbia: "between the sky" MDN, November 2006; "election promises" HN, May 2012; "first violets" HN June 2012; "spring equinox" SH v4n3; "bristled pine" SH v2n2; "dust to dust" MDN 2005; "starry sky" AHG 3.1; "autumn dusk" MDN, October 2007; "tug of her hand" THN VII.3. **p. 110–112.**

Lee Gurga, USA: "walking" AC 25; "a container" FR 37.1; "morning birdsong" NG 2.3; "two boys" *a mouse pours out* H/CP 1998. **p. 113.**

Lenard D. Moore, USA: "bullfrog's croak" MH 43.1; "hot afternoon" THN VI.9; "midday heat" DESERT STORM: A BRIEF HISTORY, LHP 1993. **p. 116.**

Linda Ashok, India: "power cut" *Significance of the Insignificant*, CYP 2012; "i am" BO 4. **p. 117.**

Lorin Ford, Australia: "on a bare twig" SHR 3; "distant thunder" HNIHC 2010; "where creek" TI 11.2; "day moon" AC 22; "winter starlight" PR 45; "harvest moon" KAHC 2012; "slow dancing" MH 44.1. **p. 114, 115.**

Madhuri Maitra, India: p. 121.

Madhuri Pillai, Australia: "afternoon stupor" CA September 2015; "second marriage" CA September 2015. **p. 117.**

Mahrukh Jal Bulsara, India: "twin bed" WHR, January 2014. **p. 121.**

Mark Harris, USA: "deep snow" R'r 11.1. **p. 126.**

marlene mountain, USA: "I can't find" R'r 7.4; "out of" R'r 7.4; "stuck in the mud" R'r 7.2; "left to itself" FR 27.2; "it's only" FR 28.2; "bit by bit" NO 4; "blackberry winter" FR 26.1; "pig and I" FR 2.3-4; "on this cold" *moment/moment moments* H/CP 1078; "thrush song" MH; "he leans" Amoskegg 1 (1980); "along with" FR 26.2. **p. 118–120.**

"between my rush to be ready" MH 29.1; "alone again" *Some of the Silence*, RMP 1999; "nothing" R'r 7.4. **p. 94–96.**

John W. Sexton, Ireland: "kitchen doodling" FR 2006; "lark" PW, summer 2009; "everywhere" HC, January 2016. **p. 97.**

Kala Ramesh, India: "receding wave" 13th MDNC 2009; "Gita chanting" AIHA 2014; "incomplete beings" R'r 13.2; "crossing river boulders" HN May 2012; "the weight" AC 31; "a world" MH 45.3; "mountain shadow" R'r 9.3; "dense fog" THN XI.2; "the ocean" MH 43.3; "I dip" MO 6; "falling blossom" PR 49; "CONGRATULATIONS" HN September 2012. **p. 98–100.**

Karen Cesar, USA: "almost autumn" MH 41.1; "evergreen" UBIHC 2014; "rolling suitcase" MH 42.3; "moon-past-its-prime" 15th MDNC 2011; "the shared smiles" VCBF 2008; "the long night" MH 45.3. **p. 102, 103.**

Karen Hoy, UK: "a new month" SP CC 2003; "Paris to Milan train" BS 19.4. **p. 101.**

Karma Tenzing Wangchuk, USA: "stone before" *Stone Buddha*, 2009; "waiting for me" PLJ Spring 1999. **p. 101.**

Kashinath Karmakar, India: "deafening rain" 18th KKHC; "after the rain" EQK, Summer 2014; "ghost town" DH Cycle 18; "window moon" SMK January 2013; "winter again" DH Cycle 18; "year's end" DH Cycle 18. **p. 104, 105.**

Kasturi Jadhav, India: "hushed night" PR 52; "vapour trail" PR 52; "sudden rain" AHG 4.2. **p. 106.**

Kathe L Palka, USA: "summer's end" PW 16.3. **p. 106.**

Kavya Kavuri, India: "spring winds" AHG 2:3 June 2013. **p. 211.**

Keiko Izawa, Japan: "poppy garden" THN VIII.2; "cold morning" THN XV.1; "pacific saury" *Modest Proposal Chapbook*, LR 2015. **p. 107.**

Kirsty Karkow, USA: "alone again" BDA book, May 2012; "mountain pass" THF Per Diem, 2013; "spring scents" *shorelines* 2007. **p. 108.**

Japanese Master: Basho: Tr. by Makoto Ueda. Taken from *Basho and His Interpreters* **p. 78.**

Japanese Master: Issa: Tr. by David G. Lanoue. **p. 79.**

jim kacian, USA: "gunshot" HGHHC 2005; "self-portrait" *Haiku 21,* MHP 2013; "dusklight" TRSMHA 2006; "between" KKHC 2008; "in a tent" *Haiku 21,* MHP 2013; "her words" R'r 11:2; "saying" FR 37.2; "from the top" SH v7n2; "somewhere" DH Cycle 18; "windy" FR 35.3; "the river" AAAAA 5; "chopping wood" NLF 26. **p. 80–82.**

Jo McInerney, Australia: "summer storm" JSA 2009; "late summer" THN XI:3; "fallen eucalypt" AC 28; "bare hills" JSA 2010. **p. 84.**

Joann Klontz, USA: "overcast" HCN 13:2. **p. 85.**

Johannes Manjrekar, India: "Sunday afternoon" FR 36.1; "monsoon clouds" HT v5n1; "rumble of thunder" TR August 2012; "night walk" KA 2013; "Republic Day" TL 2004; "traffic argument" TL 2004; "orthopaedic clinic" TL 2004; "smell of newsprint" TL 2001; "Science day" TL 2004; "women's hostel wall" TL 2003; "factory siren" TL 2003; "checkerboard paddy fields" TL 2004. **p. 86–88.**

Johannes S. H. Bjerg, Denmark: "1:1" MG 1; "beneath" NG 3.3; "wrapped" SB 3; "not mine" SB 4. **p. 89.**

John Barlow, UK: "our shadows" HPA 2007; "the wind" PR 41; "midday silence" BR 9. **p. 90.**

John Edmund Carley, UK: "slowly I search" Zip haiku. **p. 85.**

John McDonald, Scotland: "the old gossip" AHG 1.1; "birds gather" AHG 3.1; "dementia" TMR v8n2. **p. 91.**

John McManus, UK: "goldfish bowl" TI 15.1; "driftwood" TI 13.3; "backstroke" TI 11.3; "dad's house" *Inside His Time Machine,* IP 2016; "funeral home" MH 43.2; "instant coffee" AHG 1.4; "wild horses" BR 30; "cardboard boxes" *Inside His Time Machine,* IP 2016. **p. 92, 93.**

John Stevenson, USA: "jam" FR 25.2; "a man" R'r, November 2009; "the weight of poetry" MH 45.3; "winter sun" FR 23.1; "autumn leaves" MH 39.1; "three times I've said" MH 34.1; "bird" R'r 7.4; "winter morning" MH 41.2; "summer traffic" TCSM, July 1997;

2014; "at the edge" *A Snowman. Headless.* FPB 1979; "passport check" CI 1978; "tar pit" RA 2015; "only an eternal" RA 2015; "working late" HIA Japan 2003. **p. 70, 71.**

Hariharan R. "yawning wide" CA YC, January 2014. **p. 210**
Hana Masood, India: "bright blue sky" CA YC, September 2014. **p. 210.**
Hansha Teki, New Zealand: "turbulent mist" AAT, November 2014; "scented night" AAT, September 2014; "beyond me" AAT, June 2014. **p. 72.**
Harleen Kaur Sona, India: "drop by drop" MH 45.2. **p. 72.**
Harshavardhan Sushant, India: "grocery shopping" CA YC May 2014. **p. 210.**
Helen Buckingham, UK: "they search for my cervix" MH 43.2; "Chinese New Year" THN VI:9. **p. 73.**
Hema Ravi, India: "Women's Day" AHG 2.3. **p. 73.**
Hidenori Hiruta, Japan: "Mt. Fuji" SH v12n20. **p. 74.**

Ian Storr, UK: "butterflies" PR 28; "Spinning eggshells" PR 23; "Moving day" PR 23. **p. 75.**
Iliyana Stoyanova, UK: "winter morning" NHC, BHU 2015; "distant thunder" EQK June 2015. **p. 74.**
Iqra Raza, India: "long journey" MLDH, August 2015, "darkness" CA YC, January 2015. **p. 211.**

Janak Sapkota, Nepal: "an evening of gunfire" *Lights Along the Road,* BP 2005; "the rickshaw boy" *Lights Along the Road,* BP 2005; "long days of rain" *Whispers of Pine,* OR 2012. **p. 76.**
Jane Reichhold, USA: "moving into the sun" *A Dictionary of Haiku,* AHAB 1992; "a barking dog" *From the Dipper. . . Drops,* HPR, 1983; "coming home" *A Dictionary of Haiku,* AHAB 1992; "asking for a ride" *Thumbtacks on a Calendar,* HPR 1985. **p. 77.**
Jayashree Maniyil, Australia: "long distance" DHA, March 2016; "fast train" CR 24; "new moon" CR 26; "thick mist" AHG 4.2. **p. 83.**

ed markowski, USA: "temple path" & "valentine's day" & "half moon" details unknown; "Manhattan sunset" MDN 2003; "hitchhiking" MDN 2005. **p. 60.**

Emiko Miyashita, Japan: "sunset at the airport" GPS 2015. **p. 59.**

Emma Jones, USA: "sunlight" CA YC, January 2015. **p. 209.**

Emmanuel Jessie Kalusien, Nigeria: "drone attacks" AHG 2.1; "forests and trees" AHN FB, March 2016. **p. 59.**

Fay Aoyagi, USA: "who will write" *Beyond the Reach of My Chopsticks,* BWP 2011; "a hierarchy" *Beyond the Reach of My Chopsticks,* BWP 2011; "icy rain" BWHW 2008; "ants out of a hole" *In Borrowed Shoes,* BWP 2006; "a hole in my sweater" *In Borrowed Shoes,* BWP 2006; "night chill" *In Borrowed Shoes,* BWP 2006. **p. 62, 63.**

Ferris Gilli, USA: "night heat" *Shaped by the Wind,* SP 2006; "the baby reaches" MH 39.1; "spring sun" *Shaped by the Wind,* SP 2006; "more pills to take" AC 25. **p. 64.**

Francesca Dina Marie Cotta, India: "stilted forest" MLDH, August 2015. **p. 209**

Francine Banwarth, USA: "shortest day" PWLA 2014; "the river freezes" HNIHC 2011; "skipping stones" FR 33.1. **p. 61.**

Freddy Ben-Arroyo, Israel: "my wife at the piano" AHG 5.1; "memorial wall" BO 1; "alone" SB 2; "Home Sweet Home!" SB 3. **p. 65.**

G.R.LeBlanc, Canada: "Milky way" AC 33; "after the squall" NG 2010. **p. 67.**

Garima Behal, India: "que sera sera" SB 4. **p. 68.**

Gautam Nadkarni, India: "foggy night" AHG 5.1; "sitar crescendo" AHG 4.4; "after the family photo" SH Winter 2008; **p. 66.**

Geethanjali Rajan, India: "spring in her step" WHR August 2010; 'hearing aid" AHG 2.4; "clear sky" MDN 2013 "now we can talk" SBFASC 2015; "music class" WHR April 2012. **p. 68, 69.**

George Swede, Canada: "hiding somewhere" *Poetry Nippon* (1978); "canyon campfire" *Muttering Thunder* 2014; "the apple's crispness" FR

Claire Everett, UK: "winter sun" THN XIV.1; "cumulonimbus" AC 28; "fiddleheads" AHG 3.2; "scent of snow" THN XIV.3; "just-fledged light" PR 45; "butterfly dust" AC 28. **p. 50, 51.**

Cor van den Heuvel, USA: "the shadow" CP NY 1980; "through the small holes" *From the window-washer's pail* CP NY 1963; "summer breeze" TAO; "from behind me" *From dark* CP NY 1980. **p. 52.**

Cynthia Rowe, Australia: "earth hour" Creatrix 10; "last ferry home" NZPSA: Take Back Our Sky 2014; "shipping lane" 68th BFA 2014. **p. 47.**

Darrell Lindsey, USA: "prey" HNIHC 2010. **p. 54.**

David Cobb, UK: "in the bedroom mirror" *A Bowl of Sloes,* SP 1999; "a poky hotel" *A Bowl of Sloes,* SP 1999; "the journey goes on" *Palm,* EP 2002; "supper alone" *Palm,* EP 2002. **p. 53.**

David G. Lanoue, USA: "spring dawn" MH 44:3; "was that the last cicada?" BR 32; "longest day" BR 30. **p. 54.**

David McMurry, Canada: "white lilies" *Collected Haiku of David McMurray,* CPCEP, Volume 8. **p. 55.**

David Steele, UK: "evening stroll" PR 41; "behind birdsong" HS 20; "stuck to the slab" BS 11.4. **p. 56.**

Dawn Bruce, Australia: "summer sands" AM 5; "lagoon shadows" AM 5; "moreton bay fig" FRI 25. **p. 55.**

Dietmar Tauchner, Austria: "southbound" 18th KKHC 2013; "the child's eyes" *as far as i can,* RMP 2010; "speed of night" *invisible tracks,* RMP 2015; "deep inside" *as far as i can,* RMP 2010. **p. 57.**

Dimitar Anakiev, Slovenia: "El Camino del Mar" *San Francisco Haiku, 121 poems by Dimitar Anakiev,* KB 2015. **p. 58.**

Disha Upadhyay, India: "windy morning" AHG 2:3 June 2013. **p. 209.**

Don Baird, USA: "nagasaki" HNINC 2013; "folding cranes" *Haiku: the Interior and Exterior of Being,* LBP 2013; "returning" *Haiku: the Interior and Exterior of Being,* LBP, 2014. **p. 58.**

Carol Raisfeld, USA: "gossip column" WHR 5.1; "just married" SH v4n2; "breakfast together" WHR 5.1; "the palmist gasps" SH v4n2; "garage sale" SH v4n2. **p. 36, 37.**

Carole MacRury, USA: "intermission" TBDA 2010; "late afternoon" WHCDK; "heat wave" HF 2. **p. 39.**

Carolyn Hall, USA: "New Year's day" THN VIII.2; "so suddenly winter" THN VII.1; "rain-streaked windows" THN VII.3; "plum blossoms" TRSMHA 2006; "tulip buds" FR 38.1. **p. 38.**

Catherine J.S. Lee, USA: "sleet storm" STK January 2009; "paper cranes" THN XI.3; "falling leaves" TRSMHA 2011; "birth certificate" SMK July 2008. **p. 42.**

Ceya Davis, India: "soaring eagles", CA YC, May 2016. **p. 208.**

Chad Lee Robinson, USA: "prairie stream" THN VII.3; "kick by kick" NOON 3; "full moon" THN VIII.4; "the Big Dipper" HNIHC; "migrating geese" THN XIII.1; "meadowlark" FR 35.2; "my body thinner" FR 33.2; "with" AHG 3.1. **p. 40, 41.**

Charishma Navneet Gupta, India: "summer vacation" THN XV.4. **p. 43.**

Charles Trumbull, USA: "pansies" MH 33.2; "Hotel Marconi" FR 36.2; "raking into piles" *A Five-Balloon Morning*, RMP 2013; "first Christmas" *A Five-Balloon Morning* 2013; "hometown visit" *A Five-Balloon Morning* 2013. **p. 44, 45.**

Chase Gagnon, USA: "last embers" CA YC, January 2014. **p. 208.**

Chen-ou Liu, Canada: *"im-mi-grant"* 7th KHC 2014; "a deceased friend" THN 2011. **p. 43.**

Cherie Hunter Day, USA: "a splinter" MH 36.3; "looking up" THN VIII.1; "windows scratching" R'r 12.2; "cranial sutures" *apology moon,* RMP 2013; "azaleas' *apology moon,* RMP 2013. **p. 46.**

Christopher Herold, USA: "spring morning" THN XI.1; "jays at the feeder" THN VI.11; "looking" THN XIII.1; "not quite dawn" *Loose Change,* HSA Anthology 2005; "cloud shadow" *Morning Snow,* TAP 1993. **p. 48, 49.**

Ann K. Schawder, USA: "wind through" MH 45.1; "through eyes" TI 14.1. **p. 23.**

Archana Kapoor Nagpal, India: "silence of snowfall" KIHC 2014 **p. 23.**

Aruna Rao, India: "stuck thoughts" SB 1; "my fingers" MG 3; "my lies" UtB 2014. **p. 19.**

Arvinder Kaur, India: "swirling leaves" FR 37.1; "wheat stalks" *Dandelion Seeds* AP 2015; "games at twilight" IUH WH 87; "teething pup" AHG 4.1: "rickety bridge" AHG 4.1. **p. 26, 27.**

Aubrie Cox, USA: "old books" BR 26; "roadside violet" AC 26; "mating dragonflies" FR 35.2; "haiku conference" Wet Cement The Cradle 2 Anthology; "in addition" FR 36.3; "firefly flashes" FR 37.2. **p. 24, 25.**

Aviral Gupta, India: "bright sunlight" AHG 2:3 June 2013. **p. 208.**

Barbara Taylor, Australia: "crescent moon" WHR, August 2010. **p. 27.**

Barshani Gokhale, India: p. 30.

Ben Moeller-Gaa, USA: "steeping tea" THN XV:2. **p. 29.**

Beverly George, Australia: "news of a grandchild" NG 2.4; "one tiny feather" PR 47; "train tunnel" PR 22; "failing eyesight" WHCDK 2003/2004. **p. 28.**

bhavani, India: "village girls" BR 19; "winter evening" WHR, March 2008. **p. 29.**

Bill Kenney, USA: "snow in the city" MA 1; "the mirror" AHG 1.4; "awakened" THN V.15; "first date" MH 43:2; "barefoot" AC 21; "the brook flowing" MOS, autumn 2008; "deep autumn" FR 35:1; "February wind" SMK, April 2006; "late autumn" AHG 2.3. **p. 30–32.**

Billie Wilson, USA: "campfire sparks" THN XII:2. **p. 33.**

Bob Lucky, Saudi Arabia: "an old argument" FR 31.2; "the sound of bells" PW 10; "cold drizzle" WL 7; "heat wave" THN IX:3; "raking leaves" PR 48; "day moon" TI 13.2; "nothing" MH 43.1. **p. 34, 35.**

Brijesh Raj, India: "waterfall" AHG 5.2. **p. 35.**

Bruce Ross, USA: "island cottage" BR 29; "spring thaw" THN VII:3. **p. 33.**

"train whistle" PR 47; "this small ache" MH 44.1; "snowing" SP
CC 2011; "political election" HN 2.24, 2013. **p. 6, 7.**

Alexis Rotella, USA: "300 miles away" THA 2000; "Around the corner"
Between Waves RMP 2015; "ghostown" SB 2 "Mastectomy" BO 6.**p. 9.**

Alice Frampton, USA: "how many times" FR 35.2. **p. 10**

Allan Burns, USA: "half-lotus" AC 21; "climbing in shadow" THN IX.1.
p. 10

an'ya, USA: "mountain slate" TWHC 2003; "after its first flight" BHS's
J. W. Hackett Award 2001; "june breeze" THN II.12; "polarized sky"
BHS's J. W. Hackett Award 2005; "bitter cold" THN II.7; "soft breeze"
APWC 2004; "moonset" FSPHC Award 2001; "quiet cove" APWC
2002. **p. 12, 13.**

Anatoly Kudryavitsky, Ireland: "boundary stone" NG 14; "on the steps"
SHR 22. **p. 17.**

André Surridge, New Zealand: "evening breeze" JBIHA 2012; "x-ray
clinic" JRP, 2011; "lavender stalk", ESLA 2007. **p. 11.**

Angelee Deodhar, India: "rain after rain" WPHC 2015; "gibbous moon"
THN XVII. 2, 2015; "sharing an umbrella" HSA MA 2001; "rumours
of war" MH 34.2; "haiga workshop" HCN XV.3; "New Year's
bells" MDN, 2003; "water-warm boulders" MIC 1998; "in the
shadows" BR V.2; "late afternoon" THN V.7; "in the silence"
FR 21.2; "monastery" HH 133; "doing laundry" ASHN 2014.
p. 14–16.

Anita Virgil, USA: "and after such a year" *A Year Long* PP 2002; "never a
more beautiful spring" *A Year Long* PP 2002; "every window gone"
WHR March 2009; "bitter cold" THN II.6; "feet up" *A Year Long* PP
2002; "she always reads" SH 4.2; "Choking" *One Potato Two Potato
Etc* PP 1991; "*really* alone" *Summer Thunder* PP 2004; "let's take" *One
Potato Two Potato Etc* PP 1991; "speeding along" *Summer Thunder* PP
2004. **p. 20–22.**

Anitha Varma, India: "ancient banyan" THF Per Diem 2015; "stringing
beans" MH 45.2; "cicada's song" CA May 2014; "hung to dry" AHG
3.3. **p. 18.**

Iron Press; KA: First Katha Ebook of Haiku, Haibun, Senryu and Tanka Katha; KB: Kamesan Books; KP: Karakia Press; LBP: Little Buddha Press; LHP: Los Hombres Press; METP: Modern English Tanka Press; MHP: Modern Haiku Press; NZPSA: New Zealand Poetry Society Anthology; OPC: Okada Printing Co.; OR: Original Writing; PA: Partridge; PH: Press Here; PP: Peaks Press; RMP: Red Moon Press; RTB: Raku Teapot Book; SP: Snapshot Press; SPR: Saki Press; TAP: Two Autumns Press: THA: The Haiku Anthology; TRMAELH: The Red Moon Anthology of English Language Haiku; UDS: Upstate Dim Sum.

A. Jenita Annline, India: "dry leaves rise" MLDH, August 2015. **p. 206**

Aanya Singh, India: "stormy night" MLDH, August 2015. **p. 206**

Aashna Banerjee, India: "dancing without" MLDH, August 2015; "remaining still" CA YC, May 2014. **p. 207**

Aditya Ashribad, India: "still water" CA YC, January 2014. **p. 206**

Adheip Rashada, India: "shell shock" CA YC, May 2014. **p. 207**

Aditya Bahl, India: "mountain behind mountain" IUH WH 66; "dandelion" NG 17; "solving for x" AC 21; "windshield" LILJA August 2013. **p. 1**

Adjei Agei-Baah, Ghana: "roasting sun" THN XVIII.1. **p. 3**

Ajaya Mahala, India: "temple tank" DH Cycle 17; "mosquito wings" SMK May 2014; "excess dinner bill" SB 4; "family get-together" SB 2; "birdsong" TI 14.1. **p. 2.**

Aju Mukhopadhyay, India: "clouds of dust" SHR 19. **p. 8.**

Akila G, India: "morning mist" IUH WH 2015; "hundred" CR 27. **p. 8.**

Al Fogel, USA: "Basho's frog" THF Per Diem 2012. **p. 3.**

Alan Pizzarelli, USA: "far down the railroad tracks" *Paperclips* PH 2001; "the cricket cage door" *Frozen Socks* HH 2015; "only the distant peaks" *Haiku Cowboys* IB 2003; "done" *The Flea Circus* IB 1989; "squinting" *The Flea Circus* IB 1989; "reaching for" *The Flea Circus* IB 1989; "all excuses spent" *Frozen Socks* HH 2015. **p. 4, 5.**

Alan S. Bridges, USA: "sine wave" THN XVI.4. **p. 17**

Alan Summers, UK: "lime quarter" PR 13; "lullaby of rain" THN XIV.4;

TI: tinywords; TL: Temps Libre; TMR: Taj Mahal Review; TR: Troutswirl; TSC: The Sunflower Collective; TZS: The Zen Space; UN: Unfold; UtB: Under the Basho; WH: Whispers; WHR: World Haiku Review; WL: White Lotus.

Contests: AIHA: Akita International Haiku Award; APWC: American Pen Women Competition; CKK: Caribbean Kigo Kukai; EQK: European Quarterly Kukai; ESLA: E. S. Lamb Award; FSPHC: Florida State Poet's Haiku Contest; GBSC: Gerald Brady Senryu Contest; HBA: Herb Barrett Award; HGHHC: Harold G. Henderson Haiku Contest; HH: Haiku Headlines; HNIHC: HaikuNow! International Haiku Contest; HPA: Haiku Presence Award; IK: Indian Kukai; IMBA: International Matsuo Basho Award; INK: International Kukai; JBIHA: Janice Bostok International Haiku Award; JRP: Jane Reichhold Prize; JSA: Jack Stamm Award; KASIHC: Kaji Aso Studio's International Haiku Contest; KHC: Kokako Haiku Competition; KKHC: Kusamakura Haiku Competition; KIHC: The Klostar Ivanic Haiku Contest; MDNC: Mainichi Daily News Contest; NHC: National Haiku Contest; NYDD: New Year's Day Double Kukai; PIHC: Polish International Haiku Contest; PWLA: Peggy Willis Lyles Award; RHBA: R.H. Blyth Award; SMK: Shiki Monthly Kukai; SP CC: Snapshot Press Calendar Contest; STN: Shiki Tofu Kukai; SBFASC: Sonic Boom's First Annual Senryu Competition; TBDA: The Betty Drevniok Award; TWHC: The World Haiku Club; TRSMHA: The Robert Spiess Memorial Haiku Awards; UIHC: Under the Basho International Haiku Contest; VCBF: Vancouver Cherry Blossom Festival; WHCDK: World Haiku Club Double Kukai; WPHC: Wild Plum Haiku Contest.

Presses & Personal Books of Authors: AHAB: AHA Books; AP: Aesthetic Publications; AU: Author's United; BBP: Birch Brook Press; BFA: Basho Festival Anthology; BMP: Black Moss Press; BP: Bamboo Press; BWP: Blue Willow Press; CP NY: Chant Press, New York; CYP: Cyberwit Publishers; EP: Equinox Press; FPB: Fiddlehead Poetry Books; H/CP: High/Coo Press; HH: House of Haiku; HNACA: Haiku North America Conference Anthology; HP: Here Press; HPR: Humidity Productions; HSA MA: HSA Members' Anthology; HSASRA: Haiku South America Southeast Region Anthology; IB: Islet Books IHA: Indian Haiku Anthology; IP:

INDEX OF POETS AND CREDITS

Abbreviations:

Online and Print journals: AAAAA: ant ant ant ant ant; AAT: An Autumn Testament; AC: Acorn; AH: American Haiku; ASHN: Asahi Shimbun Haikuist Network; AHG: A Hundred Gourds; AHN FB: Africa Haiku Network Facebook; AL: Albatross; AM: Ambrosia; BHS: British Haiku Society; BHU: Bulgarian Haiku Union; BO: Bones; BR: Bottle Rockets; BS: Blithe Spirit; BSP: Brussels Sprout; BWHW: Blue Willow Haiku World; CA: cattails; CH: Chrysanthemum; CI: Cicada; CPCEP: Canada Project Collected Essays & Poems; CR: Creatrix; CA YC: Cattails Youth Corner; DH: DailyHaiku; DHA: Daily Haiga; DOL: dew-on-line; FB: Frozen Butterfly; FR: Frogpond; FRI: Famous Reporter Issue; FS: Frameless Sky; GI: Ginyu; GPS: Ginza Poetry Society; HC: High Coupe; HCN: Haiku Canada Newsletter; HCR: Haiku Canada Review; HF: Haiku Friends; HH: Haiku Harvest; HHA: Hyaku Haiku; HIA: Haiku International Association; HM: Haiku Magazine; HN: Haiku News; HO: Haiga Online; HR: Haiku Reality; HS: Haiku Spirit; HT: Haibun Today; HW: Haiku West; IUH WH: Issa's Untidy Hut Wednesday Haiku; KO: Ko; LILJA: Lakeview International Journal of Literature and Arts; LR: Lilliput Review; MA: Magnapoets; MAP: Mariposa; MAS: Masks; MDN: Mainichi Daily News; MH: Modern Haiku; MI: Muse India; MIC: Mirrors International Canada; MLDH: Mann Library Daily Haiku; MO: Moongarlic E-zine; MOS: Moonset; NLF: Northwest Literary Forum; NO: Noon; NG: Notes from the Gean; PJ: Prune Juice; PLJ: Parnassus Literary Journal; PR: Presence; PU: Pulse; PW: Paper Wasp; RA: Rattle; RN: Raw NerVZ; R'r: Roadrunner; SB: Sonic Boom; SBS: South by Southeast; SH: Simply Haiku; SHR: Shamrock; TAO: Terebess Asia Online; TCSM: The Christian Science Monitor; THF: The Haiku Foundation; THN: The Heron's Nest;

Tanvi Nishchal (age 16)

broken ladder
a spider weaves a web
into the web itself

Tanvi Shah (age 18)

crowded city —
a lone creeper bursts through
the concrete wall

Vividha Bhasin (age 18)

war cry —
my nephew gets ready
for the pillow fight

Sparsh Agrawal (age 10)

sunrise ...
all of the world
crowding a road

Sneha Mojumdar (age 15)

autumn morning
the vast forest doubled
by the lake

sunny morning ...
the eagle races with
its own shadow

Ritaj K. (age 14)

people everywhere
amongst them I walk
finding my own silence

Rohan Kevin Broach (age 17)

five rupees
saved in a bargain sale
my aunt feels rich

Smayan Mohanty (age 11)

pouring rain ...
the distant sound of frogs
guides me to the pond

Pruthvi Shrikaanth (age 7)

robins chirp —
two ponies nibble
apple halves

bombers' moon
once here once gone

Rishabh Jain (age 12)

trekking —
I follow the waterfall
into the woods

Lakshmi Ramaswami (age 12)

colours of Holi
a pink flower appears
as an orange one

Mihir Oak (age 18)

still mirror ...
making holes in the sky
a stone skips away

Nikita Engineer (age 18)

winter night —
every star
a different memory

Iqra Raza (age 17)

long journey
pausing for water
I drink the moon

darkness
unfolds like a song ...
granny's wordless tune

Kavya Kavuri (age 18)

spring winds
the cats try again
to tidy their fur

Hana Masood (age 18)

bright blue sky ...
shades of green
paint the fog

Hariharan R. (age 14)

yawning wide ...
I watch the leaf settle
on a bed of brown

Harshavardhan Sushant (age 18)

grocery shopping
my mother still gives me
the lightest bag

Disha Upadhyay (age 18)

windy morning —
dry leaves fall between
our conversation

Emma Jones (age 14)

sunlight
catches a ride
tide pool ripples

Francesca Dina Marie Cotta (age 18)

stilted forest
sunlight fades into mist
where trees begin

Aviral Gupta (age 18)

bright sunlight
   a shadow beside
the shadow of a friend

Ceya Davis (age 14)

soaring eagles
this winter afternoon
i wish i had wings

Chase Gagnon (age 18)

last embers
falling from the incense —
end of autumn

Aashna Banerjee (age 18)

dancing without
knowing who holds my waist …
masquerade ball

remaining still
as the sea rushes past …
I see my feet

Adheip Rashada (age 18)

shell shock
whispers in my head
louder than screams

A. Jenita Annline (age 10)

dry leaves rise
    as if winged ...
a sprinting deer

Aanya Singh (age 14)

stormy night
at the bottom of the ocean
the ship finds a home

Aditya Ashribad (age 16)

still water ...
a zebra runs away
from itself

YOUTH HAIKU

Yu Chang

writing cursive
my unspoken fear
of dancing

blinding snow
there is no need
to understand everything

back at camp
the mountain peak
still in my legs

just long enough
to leave an impression
dragonfly

Yesha Shah

sea mist —
reading the smudged words
of his last letter

third trimester —
the mother duck and I
both waddle

election fever —
my preschooler inks his
index finger

one maple leaf ...
end over end on the sand
without a trace

late autumn
the fly does not escape
my hand

Yajushi (Vinodh Marella)

walking to Ganga
a convoy of rickshaws
in tow

William J. Higginson

the clock
   chimes chimes and stops
     but the river ...

light hearted
at the very thought:
apricot blossoms

a gun in the wildflowers
one young man
more or less

deep in the arroyo
just the red handle
of a shopping cart

Vidya S. Venkatramani

muddy puddle —
the stretch of my legs
an inch too short

hushed dawn —
an egret's wing tickles
the calm lake

Vishnu P. Kapoor

winter night
even her memory
has warmth

Vinay Leo R.

salted caramel ...
the things you know
you know

autumn dusk —
the gaps I find in
grandpa's story

four-leaf clover —
still believing that
she loves me

ink puddle —
the words that
never were

Vandana Parashar

chilly morning ...
dad comes back wrapped
in the tricolour

arranged marriage
    an aunt gives me
a fairness cream

Vadivelrajan S. B.

waterfall
I search for my breath
in the biting chill

Vidur Jyoti

shared silence
all of his ninety-two years
in my hands

shrivelled leaves
some of the rain
still hanging on

another blast
and those prayers
remain unsaid

Tyrone McDonald

tunnel graffiti my brain is such a soft surface

Tzetzka Ilieva

Skype —
my parents describe
the harvest moon

did I tell you
more than you could bear?
winter hyacinth

learning to accept
what I cannot change …
ink wash painting

Tom Clausen

gone south ...
the geese, butterflies
and something in me

before sleep
laughing to myself
at myself

reading her letter —
suddenly aware of the look
on my face

rushing
        to the zendo
                to sit still

new actors —
Gandhi is shot
again

house hunt —
our dog inspects
all corners

Tito

The mountain lake today
Quietly silver
Until the sky came down ...
Bringing ravens.

Thiagarajan A.

road puddles —
umbrellas dip
at each passing car

autumn evening
I lightly touch
my son's first scribbles

cold evening —
changing my teacup
to the other cheek

Terri L. French

evening meditation
on the tao of the next wave
everything rests

not getting the whole story —
this ladybug
on my page

first snow/last snow
you say you've grown
tired of me

lingering fog
a pair of reading glasses
in every room

Svetlana Marisova

winter breeze ...
the warmth of her hand
also gone

crashing waves —
almost believing
it's forever

summer surf —
within its sound
I am sea

bitter spring —
not all ducks
are paired

Susumu Takiguchi

new potatoes —
I eat
the smell of earth too

moonless night —
I release a beetle into
deeper darkness

winter rain ...
wetting the sound
of the bugle

Susan Diridoni

step back into the fragrance our histories mingling

Susan Burch

half-empty cup
I decide I've had enough
of you

```
            y
          l   o
        l       f
      e           a
     b             s
    e               h
   h                 a
   t                 r
  n                  k
 i                     my middle finger
```

looking for love
in all the wrong places
proctologist

crowded sidewalk —
I walk into
someone's fart

winter branches
all those spaces left
for plums

Surbhi Grover

lonely road
she walks with me
wagging her tail

sprite (Claire Chatelet)

night shift —
still I put on some makeup
to lighten my darkness

Susan Constable

how easily
she changes her mind
spring rain

Mother's Day
most of our verbs
in past tense

cloudburst
the sound of raindrops
changing size

amber light
the time it takes
a leaf to fall

knife cold swimming into blue bones

Susan Antolin

night sky
one of those stars might be
the reset button

assigning my pain a number of autumn clouds

summer clouds
I pull the rope ladder up
behind me

curriculum vitae
the years
that went missing

Sreelatha Nair

from a green mound
to an even greener one ...
the sheep trail

thunderstorm
the eyes of Shiva
on her tattoo

Suhit Kelkar

mossy wall
the fear that father
put in me

Stanford M. Forrester

godless month …
i save
a pumpkin seed

when a scarecrow   isn't   the last straw

day moon —
the poem sounded better
last night

moon viewing party —
the moon
arrives late

Sondra J. Byrnes

hearing it pull
into the driveway —
my neighbor's attitude

zazen —
my knees
won't stop talking

sighing
the narrative
of snow

Sonam Chhoki

clinic waiting room —
beyond the window blinds
more wall

night border crossing —
the elephant calf holds
his mother's tail

stolen wombs —
the wind brings only dust
to the village well

Snehith Kumbla

chained dog chases the bee with its eyes

camphor flame
a mother directs her child
to pray

winter grass
the herd needs
no prodding

one by one
the fog releases
streetlights

light drizzle —
the Tanjore doll
teetering

Shobhana Kumar

blue moon
the voice
of a childhood friend

Shrikaanth Krishnamurthy

London train —
the sun skitters
off a colt

swirling leaves —
suddenly it all falls
into place

navaraatri —
every dish in the fridge
reeks of jasmine

left right left —
my canvas shoes whitened
with wet chalk

chry
sa
lis
turn
ing
in
to
some
one
i'm
not

before and after the wasteland white butterflies

wa i ti ng fo r the   ne x t   SNE   EZE

Shloka Shankar

summer heat ...
the smell of pickled garlic
from the kitchen

eraser
my mother's mistakes
no longer mine

hyphenated the twig between branches

rejection mail —
blaming it on bad
Feng Shui

typecast as a verb the moon tonight

Singh R. K.

looking for shade
under the shapeless cloud
a rag picker

Shernaz Wadia

slow day ...
the puppy plays tag
with his shadow

on wet earth
the snail draws a fine line
— a travelogue

Sheila Windsor

almost winter how thin the wind's shadow

if I spin this globe faster we merge   youandi

red balloon until there's only blue

night train
in the window just myself
not looking back

Seshu Chamarty

a rupee short …
the ice cream vendor
turns the corner

Scott Metz

only american deaths count the stars

Shabbir Shaikh

open gym
even the sun
is allowed

all the times I have sinned wildflowers

mountain pass —
weaving in and out
of a hawk's cry

munching cookies the commas in our conversation

shanty town —
the jagged edges
of moonlight

after all the things
that have gone wrong —
plum blossoms

thrush song ...
who am I
to figure you out

galaxy ...
just a lily-pond
will do

Sasa Vazic

growing silence ...
a one color world sinks
into darkness

Sanjuktaa Asopa

this puddle
what my paper boat knows
of the sea

fingernail moon
we share what's left
of the apple pie

wind-sung water
removing my sandals
to know the words

walking home
the scent of time
in acorn years

Sandip Chauhan

moonbow ...
in a grain of wheat
a farmer's song

a torn page
in a library book ...
winter rain

Sandi Pray

quiet moments —
a chickadee takes
the last one

waterfall —
a hawk's voice returns
as mist

one coyote —
the entire mountain
howls moonlight

waterfront bar
the drunk answers
a gull's call

salt free diet
somehow I knew it would
come to this

Safiyyah Patel

ruby studs —
all that remains
of nan

Samar Ghose

the
flutter
of
a
cobweb
then
the
wind
chime

city park bench
sitting briefly beside me
the midday sun

the fishermen's song parting a cobalt sky

Sandra Martyres

"z" prompts —
I am lost without
my dictionary

Sandra Simpson

spattering rain the pulse in a sparrow's throat

the moon's apostrophe —
everything I know
learned from books

photos of her father
in enemy uniform —
the taste of almonds

Rohini Gupta

vacation's end
my small black notebook
brings home the mountains

across the page
faster than my pen
a falcon's shadow

mid sentence
a furry paw bats
my pen away

Rosa Clement

orange harvest
the passing train loads
a sweet scent

endless summer …
a shadow pretending
to be a god

humid night …
a minnow polishing
stones

yesterday …
blossoms giving birth
to mirrors

Ruth Yarrow

food bank line —
a pigeon picks up crumbs
too small to see

planting peas
the earth curves under
my fingernails

Robert D. Wilson

memories
zipped up in a body bag ...
starless night

mid july ...
worker ants weaving
shadows

Ron C. Moss

old horses
days of endless rain
in their eyes

starry night …
what's left of my life
is enough

dry billabong
the colours of moonlight
in the flame trees

Roberta Beary

autumn moon
her brain a tangle
of white string

new year saké ...
the gradual dimming
of your flaws

all day long
i feel its weight
the unworn necklace

needle juniper my dna on your pocket comb

rustling leaves
it's not that hard
to be born again

blue jean patches
the sky will always belong
to my mother

empty park
two crows start
the world over

Renée Owen

dusk on the mountaintop as if I had wings

evening chill
the bottle-fed lamb
shadows a ewe

Robert Epstein

star jasmine
there's no right way
to die

zen garden
nothing
stands out

Richard Gilbert

as an and you and you and you alone in the sea

dedicated to the moon
I rise
without a decent alibi

our yesterday behind the blinds

Randy Brooks

cracked glass
over the photo
her lips parting to speak

October light
    I open my ribs
to pray

Rochelle Potkar

periwinkle sky —
the unwanted butterfly
after our breakup

traffic snarl —
I remember the song
of estuaries

train window —
a raindrop stretches
into eternity

Rebecca Lilly

Evening rain —
the downrush of day
into shadow

Fading day moon —
cool scent of clay
from the cemetery

The gravedigger staring
into space
then back at the grave

a heron's pulsing throat ...
the river thin
on the river bed

village fair ...
insects fly round and round
a tube light

temple ruins ...
an eroded Buddha
still in meditation

back from my hometown
scent of ripe mangoes
in the empty bag

carnival over …
a little girl's sandal
among footprints

sound of rain
the faded 'e' on
the internet cafe keyboard

scent of jasmine …
the night watchman
starts to whistle

daybreak …
a bicycle leans on a single tree
in the paddy field

Ramesh K.

vacation over ...
the hills we climbed
recede

starlit sky ...
I touch a turtle before
it enters the sea

rustle of palm leaves ...
fishermen play cards
in the boat's shade

clouds drift ...
the trail of a raindrop
on a dirt covered leaf

Raamesh Gowri Raghavan

silent night only the string lights Morse coding

puja flowers
swept away ... I frame you
in 12" by 8"

autumn shadow
the tall and thin man
I never was

Quamrul Hassan

cricket fever
the CEO talks batting order
with the peon

north wind —
the fisherman struggles
to light his cigarette

Rajeshwari Srinivasan

back home
I fiddle my keys
into silence

white lily —
I scrub my face
harder

Poornima Laxmeshwar

orange sunset the colour of candy painted tongues

night deepens ...
mother re-stitches
our broken dreams

Pravat Kumar Padhy

temple bell —
the lone bird adds
its cry

early morning —
the sweeper gathers
autumn wind

Peter Yovu

a falcon dives
how completely
I surround my bones

Polona Oblak

autumn wind
the slant in the handwriting
of my former self

bark
becoming whimper
becoming night

what i said
what you think i said ...
rotten plums

Peter Newton

standing in the middle of now here

yard sale
the empty fishbowl
still wet

over my thoughts the hush of pines

Purushotham Rao Ravela

spring rain ...
the tree and I
in conversation

Penny Harter

do I gather peonies
or do they gather me —
the summer garden

redwood bark —
the skin on my hand
suddenly younger

migrating butterflies
cover the names —
war memorial

Paul W. MacNeil

paddle at rest
beads of water slide
from the loon's bill

stepping stone
a hiker rests
in the river's wind

windless heat
an alligator nose
in the coot's wake

Payal A. Agarwal

between us a walkway of rain clouds

piano lesson
her braids outdo
the metronome

Indian summer
a turtle on a turtle
on a rock

summer night
we turn out all the lights
to hear the rain

before we knew its name the indigo bunting

country church
the summer smell of cotton
freshly ironed

morning twilight
fine powder on the mirror
where the moth was

I shake the vase
a bouquet of red roses
finds its shape

Peggy Willis Lyles

in spite of everything forsythia

lunch at the zoo
even among gorillas
some who sit apart

boarding call
the ripe banana flavor
of the small one's cheek

into the night
we talk of human cloning
snowflakes

rehearsing a lie
the orange brilliance
of a poppy field

with a cheap beach towel
I change the tide
on a distant planet

spring light
a scar with a beginning
and an end

deep winter
stars between the stars
I know

paul m.

daffodil shoots —
all these years
as an accountant

a line borrowed
from another poet
spring rain

bulbs dividing underground
we talk of children
we cannot have

chardonnay …
the scent of summer
in a blue-throat's song

thunderclap —
the infant's fist tightens
around a dream

fading daylight —
an oarsman's ballad
drifts ashore

Padma Thampatty

fallen leaves ...
the way the trail
takes us back

Paresh Tiwari

music box —
the ballerina held up
with Band-Aids

raked leaves ...
tossing an old tune
into the bonfire

Patricia Prime

flea market
a carton of books
I recognise as mine

riverside
the poet juggles
crab apples

in the margins
of a library book
inexpert notes

Owen Bullock

my shadow   could be anyone

Pamela A. Babusci

changing kimonos
between seasons ...
my ordinary life

i traveled a moonbeam tonight searching for you

no strength to pray ...
i place chrysanthemums
at Buddha's feet

Niranjan Navalgund

key chain the way i clingontoher

Olivier Schopfer

far from home
the rustle of willow leaves
speaks my language

end of story
you peel your tangerine
in one continuous piece

round the corner a new asphalt driveway dental scaling day

on the manuscript
the shadow of a butterfly
finishes the poem

taking a hard look
at myself from all angles —
the men's store mirrors

autumn twilight:
  the wreath on the door
    lifts in the wind

after the bell,
within the silence:
within myself

Nick Virgilio

lily:
out of the water ...
out of itself

Easter morning ...
the sermon is taking the shape
of her neighbour's hat

a poppy ...
a field of poppies!
the hills blowing with poppies!

the moon
has found it for me,
a mountain path

spring dream ...
slipping my wings
into a work shirt

Michael McClintock

having no thought
we've come to see them —
dogwoods in bloom

done for the day
my dad brings to supper
the smell of turned earth

a broken window
reflects half the moon,
half of me

the waiter interrupts
our argument on abortion —
a choice of teas

grocery shopping —
pushing my cart faster
through feminine protection

warm winter evening —
the chairs askew
after the poetry reading

crackling beach fire —
we hum in place of words
we can't recall

Michael Rehling

as if i cared or not the boolean nature of snow

pretending not to be there the whitetail and I

prayer beads
one hundred and eight times
i ask why

Michael Dylan Welch

dwindling fire —
our conversation shifts
to death

Melissa Allen

spring rain backwards until the beginning

nothing
I didn't know
before
maple
after
maple

particles decaying at the speed of lilac

the smallest bone in my body breaks winter reeds

autumn sky
only one of us
deciduous

Mark Harris

deep snow
in a dream, I find
her password in

Minal Sarosh

rippling laughter
face to face with myself
on water

Mykel Board

fiftieth birthday
standing a little closer
to the toilet

Maya Lyubenova

herring clouds
the fisherman's net
heavy with light

broken wall …
the barn and a cherry-tree
lean on each other

the b-flat
fades from her piano …
autumn wind

gamma therapy ward —
the women in the waiting room
try on wigs

Martin Lucas

the thyme-scented morning lizard's tongue flicking out

summer river —
when I'm barefoot
it's forever

morning sun
the bright white
of distant gravestones

crescent moon
the air we breathe
is propaganda

quiet sunday
the shadow of the elm
makes its round

tin soldiers
the dead and the living
in the same box

out of the haze
the dog brings back
the wrong stick

Max Verhart

falling apple —
the branch sweeps into
a new balance

autumn wind —
again the postman
passes by

Madhuri Maitra

autumn years
the salesgirl recommends
mid-rise jeans

halitosis —
the patient revives
the dentist

Mahrukh Jal Bulsara

twin bed
our baby tries
to find her space

on this cold
        spring 1
    2 night 3 4
        kittens
          wet
           5

thrush song a few days before the thrush

he leans on the gate going staying

along with wind and mud and whatever that means if anything

it's only february comes after it's only january

bit by bit a bit on 'government secrecy'

blackberry winter a touch of perspective

pig and i spring rain

marlene mountain

i can't find the time destroyed by the past

out of nowhere isn't

stuck in the mud reeds full of spaces for nymphs and me

left to itself a moon without subtitles

Linda Ashok

power cut
i count raindrops
without a miss

i am just a few vowels

Madhuri Pillai

afternoon stupor
the silent gyrations
of a skylark

second marriage
the presence of his
absent children

Lenard D. Moore

bullfrog's croak
the only night sound
the only visitor

hot afternoon
the squeak of my hands
on my daughter's coffin

midday heat
soldiers on both sides
roll up their sleeves

winter starlight
the sound of the tuning fork
goes on forever

harvest moon
the horizon between here
and hereafter

slow dancing
to Satie
the pears ripen

Lorin Ford

on a bare twig rain beads what light there is

distant thunder
   the future
      in my bones

where creek willows weave the sunlight ducklings

day moon
   the dish rag
wearing thin

Lee Gurga

walking a white parabola afternoon cicadas

a container for medical waste (3rd from the sun)

morning birdsong requiring quotation marks

two boys the last pile of dirty snow

starry sky …
wondering if I want
the answers

autumn dusk
a leaf falls into
the sound of grey

tug of her hand —
a heron one breath away
from the sky

spring equinox —
the toilet paper roll
off-center

bristled pine —
the autumn moon
has a moustache!

dust to dust
a white butterfly
bridges the gap

Laryalee (Lary) Fraser

between the sky
and the spin of the earth
this falling leaf

election promises —
I turn my tray of seedlings
toward the sun

first violets —
somewhere a river
keeps rising

Kumarendra Mallick

autumn morning —
my shadow now
has a slight hunch

day of festivity —
the flower girl sells
her smiles

Kushal Poddar

an ancient wick
casts shadows on the wall —
Mother's soft humming

Kirsty Karkow

alone again
the last raspberry
sharp on my tongue

mountain pass
we rebuild the cairn
to fit the wind

spring scents
the dog and I walk
through different worlds

Keiko Izawa

poppy garden ...
in and out of the flowers
the child's red cap

cold morning
the sound of pouring tea
fills the room

pacific saury —
the scent of ocean
in my dish washer

Kasturi Jhadav

hushed night ...
all the universe
in a frog's croak

vapour trail
the sky divides
its share of birds

sudden rain
the fortune-teller runs
for cover

Kathe L. Palka

summer's end —
dust gathering on books
I planned to read

window moon —
an imperfect circle
in a perfect square

winter again —
somehow a coriander leaf
inside my wallet

year's end —
only the sound of mouse clicks
from every desk

Kashinath Karmakar

deafening rain —
to think it has no sound
of its own

after the rain
in each hanging droplet
the world upside down

ghost town —
the sound of army boots
from alley to alley

moon-past-its-prime
yesterday's rice bowl too
imperfectly round

the shared smiles
of passing strangers ...
cherry blossoms

the long night ...
an old woman's loneliness
follows me home

Karen Cesar

almost autumn so many holes to another universe

evergreen touching evergreen sound now water now wind

rolling suitcase ...
the homeless woman passes
as one of us

Karen Hoy

a new month —
different seeds
on the spaniel's ears

Paris to Milan train
the baby cries
in every language

Karma Tenzing Wangchuk

stone before stone Buddha

waiting for me
to give it life —
my death poem

the ocean in a raindrop inside my womb a heart

I dip my feet
in a river the river
joins the sea

falling blossom ...
the breath between what was
and what will be

CONGRATULATIONS
you've won $5,000,000.00 USD:
I count the zeros

the weight
weighs itself out ...
falling leaves

a world beyond the word I scribble my name

mountain shadow robs the tree of its

dense fog
the train evaporates
into a distant horn

Kala Ramesh

receding wave ...
crab holes breathe
the milky way

Gita chanting
  birds become
the ellipsis

incomplete beings
  you and I
complete the city

crossing river boulders
  the moon becomes whole again

John W. Sexton

kitchen doodling
I make her a necklace
of apple skin

lark
bringing pieces
of sky in its song

everywhere
in the coal sack
fossilized starlight

summer traffic
my shadow rides up
a stranger's neck

between my rush to be ready
and her arrival —
a space

alone again
making an event
of a sandwich

nothing matters how green it gets

autumn leaves
another day of money
changing hands

three times I've said
"your husband . . ."
now we can just talk

bird me catch me

winter morning —
scribbles on a scratch pad
get the ink flowing

John Stevenson

jampackedelevatoreverybuttonpushed

a man in a crowd in a man

the weight of poetry
where snow
begins to fall

winter sun
a stranger makes room
without looking

funeral home feeling nothing about feeling nothing

instant coffee
my neighbour and me
set the world to rights

wild horses I forget about being

cardboard boxes
my son turns himself
into a robot

John McManus

goldfish bowl
my daughter asks
if I'm going to die

driftwood
we discuss
our origins

backstroke the sound of my mother's womb

dad's house
I unbutton a shirt
that no longer fits

John McDonald

the old gossip —
her grave
nearest the squeaky gate

birds gather —
the gravedigger
slowly disappearing

dementia —
he sits in the midst
of a scattered jigsaw

John Barlow

our shadows holding hands the width of the stubble field

the wind being farmed the wind that isn't

midday silence
sun-highlighted fingerprints
on the acoustic

Johannes S. H. Bjerg

1:1 ratio the white pills and dusk

beneath her voice a timbre of something crumbling

wrapped in a blackbird sunrise remains a legend

not mine the space I can hold in one hand

Science day
the programme begins
with rituals

women's hostel wall
a langur rests an arm
on the barbed wire

factory siren
the myna drops its twig
for a worm

checkerboard paddy fields
two squares
without herons

Republic Day
the fruit seller's flag is stuck
into a banana

traffic argument
the camel's sneer
is impartial

orthopaedic clinic
a three-legged chair
outside the entrance

smell of newsprint
twenty beauty queens
with one smile

Johannes Manjrekar

Sunday afternoon
the silence heavier
after the barking

monsoon clouds
crow silhouettes complete
the leafless tree

rumble of thunder
a sunbird comes darting
through the wire fence

night walk
i slow down
near the jasmine bush

Joann Klontz

overcast
I bother the blackbirds
for a glimpse of red

John Edmund Carley

slowly I search   a field of flowers
   finding nothing   but beauty

my bedroom wall I conjure up a snowdrop

Jo McInerney

summer storm
the windscreen wipers
slice our silence

late summer
crossing the memory
of a stream

fallen eucalypt ...
the scent
cut into stove lengths

bare hills
the horizon looped
between post and wire

Jayashree Maniyil

long distance
the non-stop call
of a blackbird

fast train —
the cry of seagulls
before and after

new moon ...
the same old bills
at the new address

thick mist
the sound of a cowbell
leaving the shed

somewhere becoming rain becoming somewhere

windy day i think in music

the river
the river makes
of the moon

chopping wood —
someone does the same
a moment later

in a tent in the rain i become a climate

her words inventing another piece of my mind

saying Ō to all things snow

from the top floor several weathers in the big city

jim kacian

gunshot the length of the lake

self-portrait some truth to the smudges

dusklight —
I read her poem
differently

between statues the rest of history

Kobayashi Issa
Tr. David G. Lanoue

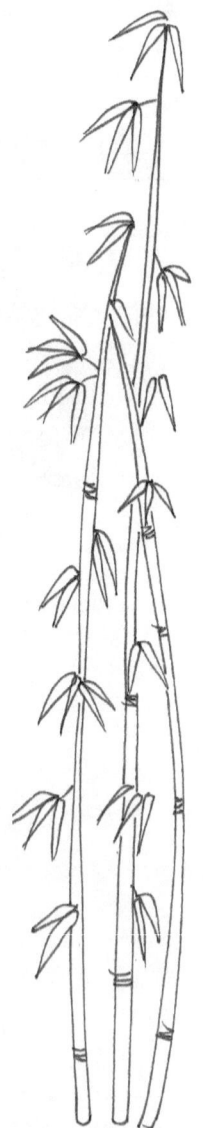

little snail
inch by inch, climb
Mount Fuji!

my dead mother —
every time I see the ocean
every time …

stitching together
the short summer night …
singing frogs

this world
is a dewdrop world
yes … but …

JAPANESE MASTERS

Matsuo Basho
Tr. Makoto Ueda

on a bare branch
a crow has alighted ...
autumn nightfall

twilight dawn
a whitefish, with an inch
of whiteness

the stillness —
seeping into the rocks
cicadas' screech

the sea darkens
a wild duck's call
faintly white

Jane Reichhold

moving into the sun
the pony takes with him
some mountain shadow

a barking dog
little bits of night
                breaking off

coming home
flower
      by
         flower

asking for a ride
his answer dotted
with music

Janak Sapkota

an evening of gunfire
a wild goose
searches for its mate

the rickshaw boy
waiting for a customer
falls asleep

long days of rain —
the gurgle of frogs ripens
my little rice field

Ian Storr

butterflies among the brambles in a field of bricks

Spinning eggshells
on the kitchen counter
the night wind

Moving day
I show the new owner the knack
with the back door

Hidenori Hiruta

Mt. Fuji
rising in a field of clouds
summer dawn

Iliyana Stoyanova

winter morning
a child draws snowflakes
with his nose

distant thunder
a line of footprints still
escaping the waves

Helen Buckingham

they search for my cervix
orchids on the ceiling

Chinese New Year —
daring to call my sister
a monkey

Hema Ravi

Women's Day
eating
leftovers

Hansha Teki

turbulent mist —
all that is to become
lost in becoming

scented night —
a moth enters the hush
between stars

beyond me
the echo of a song
once me

Harleen Kaur Sona

drop by drop
it becomes a river —
sound of rain

passport check
my shadow waits
across the border

tar pit an urge for immortality

only an eternal present jackhammer

working late
i meet my loneliness
in the long hallway

George Swede

hiding somewhere in this room a good idea

canyon campfire
our shadows as large as
the lives we wished for

the apple's crispness ...
i put an x thru
the new poem

at the edge of the precipice   I grow logical

clear sky —
the vendor sells clouds
of cotton candy

now we can talk
of what might have been —
menopause

music class —
a chorus of sneezes
after the rains

Garima Behal

que sera sera
i frown over my
daily horoscope

Geethanjali Rajan

spring in her step —
she skips
to the next sunspot

hearing aid —
grandma now complains
about the squirrels

G.R. LeBlanc

Milky Way
the numerical perfection
of a snail's shell

after the squall
the tinkling concerto
of sailboats

Gautam Nadkarni

foggy night
    an auto-rickshaw
shredding the silence

sitar crescendo skyward rising of a lark

after the family photo
mom removes
her dentures

confused drunk ...
not knowing when to zig
and when to zag

video conference —
with great care she selects
the right lip gloss

Freddy Ben-Arroyo

my wife at the piano —
why so many notes
Chopin?

memorial wall my son's name carved on sirens

alone on the lawn sprinkler psits..psits..psits..psits

Home Sweet Home!
back from a break in London
I keep to the right

Ferris Gilli

night heat
nothing moves
but the gecko's eyes

the baby reaches
for a leaf shadow ...
our talk of old loves

spring sun
the blind girl asks
the color of my hair

more pills to take —
the living half of the dogwood
in full bloom

ants out of a hole —
when did I stop playing
the red toy piano?

a hole in my sweater
I ask him one more time
what he meant

night chill
rearranging the order
of canned soups

Fay Aoyagi

who will write
my obituary?
winter persimmon

a hierarchy of apples in the moonlight

icy rain
at the bottom of the lake
a door to yesterday

Francine Banwarth

shortest day
spare parts
in a box

the river freezes ...
silence is also
an answer

skipping stones ...
I remember
what he forgot

ed markowski

temple path
the dust i kick up
sticks to me

valentine's day
we pass the lip balm
back and forth

half moon
I lie to right
a wrong

Manhattan sunset
the street magician's
first trick

hitchhiking
an orange moth fills
the emptiness of Texas

Emiko Miyashita

sunset at the airport —
empty postcards
on my lap

Emmanuel Jessie Kalusian

drone attacks
my child goes to bed
with leftover stones

my father's shoes
no matter how you fix it
an open mouth

forest and trees —
i visit my ancestors
with no dry gin

Dimitar Anakiev

El Camino del Mar
where the small bird and I
watch the ocean

Don Baird

nagasaki ...
in her belly, the sound
of unopened mail

folding cranes ...
an autumn gust lifts
one away

returning
the sound of gunfire
returning

Dietmar Tauchner

southbound birds the loop of identity

the child's eyes green beyond the barbed wire

speed of night
somewhere beyond big bang
i am a unicorn

deep inside you no more war

David Steele

evening stroll
just enough drizzle
to give each drain a voice

behind birdsong
         a distant sound of mowing
                  comes and goes

stuck to the slab
the   i
of the frozen f sh

David McMurray

White lilies
the feeding tube
removed

Dawn Bruce

summer sands
shadows gathered
in heel marks

lagoon shadows
slide across the ceiling
new lover

moreton bay fig
spreads a hundred years of shade

David G. Lanoue

spring dawn
I put on
my gender

was that the last cicada?
September
shadows

longest day
deleting the dead
from my phone

Darrell Lindsey

prey on the cave wall an arrow's unfinished flight

David Cobb

in the bedroom mirror
the old slow bowler
bowls at himself

a poky hotel
no room for my shadow
to unpack

the journey goes on
I squeeze just that bit higher
up the toothpaste tube

supper alone
the chance to whistle
through my macaroni

Cor van den Heuvel

the shadow in the folded napkin

through the small holes
in the mailbox
sunlight on a blue stamp

summer breeze
a ladder leans against
the half-painted house

from behind me
the shadow of the ticket-taker
comes down the aisle

scent of snow
unable to recall
my father's voice

just-fledged light
chips of wren song
from the log pile

butterfly dust ...
the question I never
dared to ask

Claire Everett

winter sun ...
the soft flicker of waxwings
in the firethorn

cumulonimbus
the egret preens deeper
into its breast

fiddleheads uncoiling all the time in the world

looking
not looking
road kill

not quite dawn
someone stops trying
to start a car

cloud shadow
long enough
to close the poppies

Christopher Herold

spring morning
the art of walking beside
someone much older

jays at the feeder —
I remember names
of high school bullies

Cynthia Rowe

earth hour …
threading beads
by moonlight

last ferry home
the night empty
of stars

shipping lane
the mist swollen
with foghorn

Cherie Hunter Day

a splinter comes out whole   winter moon

looking up
rules of punctuation —
the green hills

window scratching the sheen of a failed wing

cranial sutures
the continents no longer
fit together

azaleas as afterthought as afterword

first Christmas
without my mother
without my childhood

hometown visit
fine sand in the doorways
of vacant storefronts

Charles Trumbull

pansies   we smile back

Hotel Marconi
in every room
free wi-fi

raking into piles
leaves from a tree
I climbed as a boy

Charishma Navneet Gupta

summer vacation —
grandmother's stories stretch
from sunrise to sunset

Chen-ou Liu

im-mi-grant ...
the way English tastes
on my tongue

a deceased friend
taps me on the shoulder —
plum blossoms falling

Catherine J. S. Lee

sleet storm
her needles clicking out
a red mitten

paper cranes
the yellow birch
unpleats its leaves

falling leaves
the clang of horseshoes
in the crisp air

birth certificate:
the name of the father
he never knew

meadowlark —
all you'll ever need to know
about sunrise

my body thinner these days I hear more of the wind

with
what
light
is
left
the
dish
rag
drip
ping

Chad Lee Robinson

prairie stream —
what I know about mountains
in these small stones

kick by kick the stone's shadow evolving

full moon —
all our sounds
are vowels

the Big Dipper —
rows of corn connect
farm to farm

migrating geese —
the things we thought we needed
darken the garage

Carole MacRury

intermission —
a fly on the piano
walks a full scale

late afternoon —
the fullness
of the cow's udder

heat wave
the horse blinks away
a gnat's life

Carolyn Hall

New Year's Day
the center of the chocolate
not what I expected

so suddenly winter
baby teeth at the bottom
of the button jar

rain-streaked windows
how to paint
the finch's song

plum blossoms
I make plans
for my ashes

tulip bulbs —
that first piano recital
still in my fingers

breakfast together —
the silence about things
that matter

the palmist gasps
then asks to be paid
in advance

garage sale —
the flowered couch on which
I became a woman

Carol Raisfeld

gossip column —
only the ink remains
unsmeared

just married ...
eating the oysters
he used to hate

raking leaves
the wind and I
take turns

day moon even here I'm somewhere else

nothing in the window is everything

Brijesh Raj

waterfall ...
the rock-face overrun
by tiny red crabs

Bob Lucky

an old argument —
scraping the burned rice
out of the pot

the sound of bells
wishing I believed
in something

cold drizzle —
a goodnight kiss
on my bald spot

heat wave
the mailman fans himself
with my bills

Billie Wilson

campfire sparks —
someone outside the circle
starts another song

Bruce Ross

island cottage
the daisies growing
where they want to

spring thaw
a new spider
in the mailbox

deep autumn
knowing there is nowhere
I have to be

February wind
I want to believe
the crocus

late autumn
my fear
of falling

awakened
by moonlight
an old regret

first date
the way she pronounces
Van Gogh

barefoot
the earth
pushes back

the brook flowing
through my shadow
morning chill

Barshani Gokhale

size zero ...
the roundness
of a doughnut

Bill Kenney

snow in the city
nobody home
in the cardboard box

the mirror
when no one is looking
first light

Ben Moeller-Gaa

steeping tea
the time it takes to lose a street
to snow

bhavani

village girls
         their eyes
louder than their voices

winter evening …
the roadside chaiwala
makes more money

Beverley George

news of a grandchild
I touch a furled bud
in spring rain

one tiny feather
all the colours of the bird
weightless on my palm

train tunnel —
the sudden intimacy
of mirrored faces

failing eyesight —
we sing only the carols
we know by heart

teething pup —
the tattered remains of
The Satanic Verses

rickety bridge —
the langur's leap
from fog to fog

Barbara Taylor

crescent moon
the turning point
of my mind

Arvinder Kaur

swirling leaves —
a scribble of starlings
in the evening sky

wheat stalks —
someone calls me
by a forgotten name

games at twilight ...
I am given away
by my shadow

haiku conference —
I'm everyone's
Granddaughter

in addition
to what's been said
kitten scratches

firefly flashes   the distance   of narrative

Aubrie Cox

old books
without their jackets
summer rain

roadside violet
all the places
I've yet to go

mating dragonflies —
my overuse
of dashes

Ann K. Schwader

wind through the stars my bones

through eyes of rain leaf light

Archana Kapoor Nagpal

silence of snowfall —
tinkling temple bells
in the valley

really alone:
an itch on my back
I can't reach
(s)

"Let's take our picture!"
    she says every time,
the one who's photogenic.
(s)

speeding along the awning's edge
raiiiiiiiiiiiiiiiiiiiiiiiiiiiiiiiiiiiiiiiiiiiiiiiiiiiiiin

feet up
toes spread wide
I catch
8 tiny summer breezes

she always reads the film critics
to find out
what she thinks
(s)

Choking,
    he keeps explaining
how he choked.
(s)

Anita Virgil

and after such a year
the first crocus
in its usual place

never a more beautiful spring
in which to be
ill

every window gone
the old house takes in all
the spring wind

bitter cold
the distant train pulls
some of my night with it

Aruna Rao

stuck thoughts
the pencil dragonflies
between my fingers

my fingers dial a spider

my lies
grow bigger and bigger:
how the whale evolved

Anitha Varma

ancient banyan ...
an owl shakes the night
off its feathers

stringing beans ...
a scrap of twilit sky
through the window

cicada's song ...
I listen to the sounds
of bamboo growing

hung to dry a cleaner sunshine after the rain

Anatoly Kudryavitsky

boundary stone
the nettles
pause

on the steps
of the Freedom Memorial,
a discarded snake skin

Alan S. Bridges

sine wave
a purple finch
does the math

late afternoon
up to its belly in hyacinths
a pregnant buffalo

in the silence
of the zendo
my stomach growls

monastery ...
rising above the plainchant
a warbler's half note

doing laundry
at the river's edge
the flow of gossip

haiga workshop
in the down stroke of the brush
the sound of rain

New Year's bells
all the way to the horizon
frosted cabbages

water warm boulders
so smooth now
against callused feet

in the shadows
of a field of sunflowers
a whitewashed shrine

Angelee Deodhar

rain after rain —
just one magnolia offered
to the stone Jizo

gibbous moon
my ear on the curve
of her belly

sharing an umbrella
your wet left shoulder
my right one

rumours of war
up into a darkening sky
- a child's newsprint kite

bitter cold
a juniper berry parts
the jay's beak

soft breeze
a bee's stinger lifts
in the air

moonset
a sudden vastness
between stars

quiet cove
the sound of crow's feet
cracking thin ice

an'ya

mountain slate
the color of the sound
spring rain makes

after its first flight
the young gerfalcon's talons
tighter on my glove

june breeze
a hole in the cloud
mends itself

polarized sky
the mixed melodies
of twilight birds

André Surridge

evening breeze
a flag releases
its stars

x-ray clinic
the orange bones
of a willow

lavender stalk
the weight of one
white butterfly

Alice Frampton

how many times
do I have to tell you
meteor shower

Allan Burns

half-lotus ...
the slow degrees
of dusk

climbing in shadow —
the canyon rim
brightly lit

Alexis Rotella

300 miles away —
my father makes sure
I hear him sigh

Around the corner
summer waits
in a yellow dress

ghostownothere

Mastectomy
she wakes up
alone

Aju Mukhopadhyay

clouds of dust
trailing the bus —
village path

Akila G.

morning mist ...
a stray sunbeam carves
the mountain

hundred —
Grandpa claims to be older
than the banyan tree

this small ache and all the rain too robinsong

snowing
through the blizzard
particles of me

political election
my application to be
a) human

after Fay Aoyagi

Alan Summers

lime quarter
an ice cube collapses
over jazz

lullaby of rain
another pinch of saffron
in the pumpkin soup

train whistle
a blackbird hops
along its notes

squinting
to read the sign
   "optician"

reaching for
   the wind-up toy
      it rides off the table

all excuses spent
i tell my wife
about my alien abduction

Alan Pizzarelli

far down the railroad tracks
   the brakeman's lantern
      gets lost among the fireflies

the cricket cage door
left open   starry night

only the distant peaks
between the horse's ears

done
the shoeshine boy
snaps his rag

Ajaya Mahala

temple tank —
near the stone bull
a real bull

mosquito wings
the colour of evening
so thin

excess dinner bill
she picks up
five more toothpicks

fAmiLy gEt-tOgetHer

bird song —
an incomplete sentence
each time

Al Fogel

Basho's frog ...
four hundred years
of ripples

Adjei Agyei-Baah

egrets in formation
a young one
breaks the rule

roasting sun
the egret's measured steps
in buffalo shadow

Aditya Bahl

mountain behind mountain behind mountain
petals of a rose

dandelion: dos & don'ts

solving for x ...
the trees you dream
have no leaves

windshield wiping crows in flight

HAIKU

Geeta Dharmarajan and jim kacian came to our rescue with their expertise and deep knowledge of publishing. Thanks to Rohini Gupta for helping us with the contract. Our thanks to Vishal Soni and Chaitali Nachnekar of Vishwakarma Publications for the many ways in which they assisted us.

Last but not least, we are grateful to all our families for their support, as this project took almost 18 months of our 'mind-space' in the making. And finally, to each and every poet who graciously gave us their ku for *Naad Anunaad*, our sincerest pranaams.

Kala Ramesh
Sanjuktaa Asopa
& Shloka Shankar

of her hand; it is uncanny at times to realise that we're culturally so different, and yet so alike in our thinking.

My thanks to Shernaz Wadia for helping me out at the start of this project. My deepest gratitude to Sanjuktaa and Shloka for the work they put in as editors. In this collection, we have a staggering number of haiku from poets from over 26 countries, including 31 haiku from 29 poets under 18 years of age. We believe this is the first haiku anthology to include so many haiku from such young poets.

We thank young Dishika Iyer and Ramesh Anand for proofreading the publication credits. The meticulous way in which Ramesh went about checking the credits is truly praiseworthy. Thanks to Uedo Makoto and David Lanoue, we could include the Japanese masters, Basho and Issa, in translation. Our special thanks also to the family members of Laryalee Fraser, John Carley, Martin Lucas, Yajushi, and William Higginson and to the estate of Nick Virgilio for granting us permission to reprint their ku. Special mention goes to Anita Virgil, for the interesting discussions we had on senryu.

We drew a blank when it came to choosing an illustration for *Naad Anunaad,* since the title is abstract. Alaka Yeravadekar's beautiful Water Lilies in ink and watercolours, which Geeta Dharmarajan turned into a beautiful cover, was a godsend for us! Thanks to Namrata Deshpande and Shipra Dutta for the ink drawings and calligraphy.

ACKNOWLEDGEMENTS

We, Sanjuktaa, Shloka, and I, are deeply grateful and would like to thank all the poets and friends in the haiku world, and outside, who have helped bring *Naad Anunaad* to fruition.

Our sincere thanks to Alan Summers, Carolyn Hall, Claire Everett, Ferris Gilli, Francine Banwarth, Helen Buckingham, Rebecca Lilly, jim kacian, Owen Bullock, Ramesh Anand, and Sanjuktaa Asopa for suggesting the haiku of poets that we otherwise would have missed.

The Foreword intrinsically and naturally surfaced from what I know, which is Indian, and, at each turn, I found to my surprise that it had a lot of similarities with Japanese perspectives and aesthetics. Richard Gilbert encouraged me in my finds and his detailed feedback gave me a Himalayan eagle's eye-view of haiku as it is practised in Japan today.

Throughout the making of *Naad Anunaad*, Jenny Angyal's insightful advice, reinforcement, feedback, edits, and proofreading were invaluable. She reads my mind like the back

lotus viewing
    the flowering
within

In India, as elsewhere, when we go on mountain treks, we rely on the locals to know the wind, the rain, the soil, and even how big the moon is going to be when it rises ... and if we contemplate, reflect, and explore this knowing through haiku, we may fine-tune the art of coming closer to mother nature and being attentive to her needs just as these locals have always been. We are hoping *Naad Anunaad* will draw a new generation of readers and authors into the kind of intimacy with nature that our grandmothers enjoyed. Nature here does not mean just the hills, rivers, and forests — it includes cities and the life we live, our day-to-day activities, taking in stride the agonies, failures, successes, and idiosyncrasies intertwined with the natural world.

We welcome you warmly to this anthology and to the world of haiku in all its various avatars.

Kala Ramesh
June 2016
Pune

'vast is the ocean' Rig Veda 1.3.12, Tr by Pandit Satyakam Vidyalankar.
'beej' – seed.
gate is pronounced as "gah-tay."
'lotus viewing' and 'amavasya' by Kala Ramesh.
A small section of the introduction was first published in *A Hundred Gourds 2:3,* under the title 'Haiku in India.'

As much as we would have loved to show our young and new readers the differences in treatment and texture between haiku and senryu, we decided that after a year of work and with so many beautiful poems, *Naad Anunaad* should not end in arguments and controversies. We would rather have people read and enjoy the anthology for the poems, and not for the labels we give them. (Just a quick note here: Anita Virgil wanted her senryu to be marked with an (s) and we have done so out of respect for her passion for this genre.)

In this anthology, we have included haiku written by school students and undergrads. It is great fun to work with children, for the very name 'haiku' fascinates them! Curiosity is one thing kids have in abundance, and if you catch that, you tap into some great creativity and talent. Except for a few youngsters, most of those under the age of 16 are from India, and took a one-hour workshop at the Bookaroo Children's Literary Festival or a two-day workshop in various cities as part of *I Love Reading*, a Katha Initiative, in collaboration with the school students of the Central Board of Secondary Education. Students between the ages of 16 and 18 are from the Symbiosis International University, Pune. As an elective in their liberal arts graduation programme, they chose a 60-hour, four-month course in Japanese short forms of poetry, including haiku/senryu, tanka, haibun/tanka prose, renku, and haiga.

Haiku is all about relating to the essence of being, a technique which is not easy to teach, but something that has been done to perfection by many authors.

transported to Japan, became "Zen." Dhyana yoga was taken by Bodhidharma from Kancheepuram, South India, to China, from where it penetrated into Japan. The famous 'silk route' gave a further fillip to all interactions as commerce flourished, bringing cultural trade closer among the civilizations along its network. That Indian culture and religious beliefs criss-crossed and travelled to East Asian countries is now undisputed.

While we may not yet have an Indian '*Saijiki*' — the Japanese reference book of kigo words — as the Indian haiku tradition grows, perhaps someone will take on the task of building an Indian Saijiki, which would be difficult, but doable.

To be kept alive, tradition needs to be a flowing river, for stagnant water stinks. The multifaceted haiku has been changing its face rapidly over the years. It's difficult to keep pace with it. Halfway through our selection process for this anthology, we realized that poets were submitting haiku, border-line haiku, senryu, so-called gendai (modern) haiku and the new ku. They have all found a place in *Naad Anunaad*, because we, as editors, loved them. We debated whether to include all these forms under the generic term haiku, or whether we needed to show their differences, especially in the case of haiku and senryu. Grouping them together somehow seemed to blur the distinctive aroma (or *rasa*, as we say in India) of each genre, but somehow we had to decide. Our minds played devil's advocate and kept asking *why not?* And our hearts kept saying, *why divide, why build walls.*

in the south, which is close to the equator, to the snow-tipped Himalayas in the north, and from the Arabian Sea on the western coast to the Bay of Bengal in the east; we have all the seasonal changes that anyone can think of!

An interesting and equally important point to be noted is the relationship between India and East Asia. The idea that in Japan, and especially in Zen, man was considered a part of nature is very close to Indian thought originating in Vedic times.

In Hindu philosophy, and more so in our cultural memory, the concept known as *Advaita* — non-duality, the oneness of consciousness — is deeply entrenched. Yes, we are that one pulsating consciousness — that blade of grass, that mountain, you and I are all one pulsating consciousness. The Sanskrit pronouncement '*Tat Tvam Asi*' (you are that) — appears in the Chandogya Upanishad, which was composed in the earlier part of the first millennium BCE. One of the oldest Upanishads, it forms the basis for the whole theory of oneness and non-dualism. But this 'being one' is the last and final step in the realisation of the goal. Caught up as we are in this mundane rat race called life, all of us are constantly swayed by dualisms: sadness / joy, weakness / strength, success / failure, and so on. These dualisms are what we all experience on a day-to-day basis; these dualisms are what colour us and give us our persona — and frankly, they are what makes for a good haiku.

It is said that '*dhyaan*' (dhyana yoga or meditative absorption), when spread to China became '*cha'an,*' which, when

May, and not on the day he was born according to the western calendar ('*Poornima*' means full moon).

Known as *amavasya,* the no-moon day or the first night of the first quarter of the lunar month, holds great significance — for the moon is on the way to its full potential, symbolizing the path from darkness to light, from ignorance to enlightenment.

amavasya ...
the river flows on sounds
the river makes

Incidentally, most of our mothers would trim our hair during the period immediately after amavasya, for they firmly believed growth happened only with the waxing moon.

India is primarily an agricultural country, dependent on the seasonal rains referred to as the monsoon. There are more than 100 names to describe rain throughout India. For example, *hatheetsa pous* is a Marathi phrase meaning 'elephant rains,' which occur during the ten-day Ganesh festival. *Hathi* means elephant and Lord Ganesh has an elephant face. As the name suggests, *hatheetsa pous* stands for the big drops of rain we have at the end of the monsoon, from the end of August into September on the west coast. In Chennai, which lies on the east coast, the summer heat is so severe that September is known as the month of *mann karayum mazhai* (mud dissolving rain), and *ponn urukkum veyyil* (gold melting heat.)

India is a vast country, stretching from the tip of the peninsula a spring raga, where, against a familiar backdrop, the musicians create the raga, for, unlike western classical music, which is mostly scored, Indian raga music is on the spot improvisation. In haiku, we do something similar when using the seasonal reference as a backdrop.

Even though contemporary English-language haiku does not follow the 5/7/5 sound-unit structure anymore, we would like to bring into focus some relevant points regarding the odd-number rhythm in haiku. It is interesting to observe that this mighty Buddhist mantra from the Heart Sutra, recited all over the world, has a count of seventeen syllables:

gate gate parāgate parāsamgate bodhi svāhā *

Indian classical music employs extensive use of 3, 4, 5, 7, and 9 rhythmic-beat cycles. A student spends years in training to master these rhythmic cycles, known as *tala*. We see musicians, percussionists, and classical dancers weave effortlessly through the odd-number beat cycles of 3, 5, and 7.

We still follow the lunisolar calendars for all festivals and important occasions. Dates for marriages, housewarming ceremonies, ancestors' death ceremonies, and many other occasions are based on the waxing and waning of the moon. In many traditional homes, birthdays are still celebrated on two days — one according to the Indian (lunisolar) calendar and the other according to the Western calendar. For instance, 'Buddha Poornima' (Buddha's birthday) is a national holiday and is celebrated when the full moon occurs in the month of

Prize in 2002, and shared the prize money of one million yen with the American poet, Cor van den Heuvel. Haiku in regional languages such as Hindi, Marathi, Telugu, Tamil, and Malayalam, to just name a few, is gathering momentum.

We would like to draw a parallel between haiku and the Indian aesthetic tools which are used extensively in our raga music, dance, poetry, drama, painting, and other branches of fine arts.

Seasonal reference, known as *kigo* in Japanese, is pivotal to haiku. Indian miniature paintings, especially the '*Ragamala paintings*,' known for depicting seasons, started in the 16th and 17th centuries. They stand as a classical example of the amalgamation of art, poetry, and classical music in medieval India.

Indian classical music employs seasonal references extensively in its own way. Called the theory of time and seasonal ragas, it has developed into a science and is today followed meticulously by vocalists and instrumentalists. Some prominent musicians even use seasonal ragas to heal people suffering from minor ailments.

If a person, who is totally unfamiliar with Indian classical music, walks into a concert and the singer begins with the raga *Puriya Dhanashree*, she may just listen, for to her it would be just another raga. But we know that raga *Puriya Dhanashree* is a twilight melody, when birds return home, a time sacred for dhyaan or meditation. Building on this cultural memory in the listener's mind, the musician erects an edifice. Take raga *Basant*,

NAAD ANUNAAD

medicine native to India, and create a sense of harmony with the natural world.

Add to all this the art of suggestion, the striking feature of haiku — understanding what needs to be said and what needs to be held back; what to regard as redundant and what to reveal as important when writing a haiku. Knowing and understanding this balance is the secret behind this art form. What is generally known as 'negative' space in art can be viewed as 'positive' space pulsating with life when it comes to haiku. It opens doors and windows; it opens up dialogues in readers' minds. To learn the knack of leaving things unsaid — which leads us to what the Japanese call *ma*, the void between and around things, unclutteredness — brings the distilled essence to the surface. All this is deftly dealt with in a few words, perhaps in the time even of a single breath. The famous Bharatanatyam dancer, Rukmini Devi Arundale, once said that *Abhinaya* in dance — the rendering of emotions through body postures and facial expressions — needs to be mere suggestion; anything more becomes drama. So it would not be too grand a stretch to say that with haiku, we are touching our roots.

Haiku as a poetic form fascinated both Rabindranath Tagore and Subramanya Bharathi — revered poets from Bengal and Tamil Nadu — at the beginning of the last century. A recent phenomenon in our haiku landscape has been Satya Bhushan Verma, Professor Emeritus at Jawaharlal Nehru University. He was chosen for the Masaoka Shiki International Haiku

India is slowly, but surely, waking up to the beauty of haiku. The reasons are not far to seek. Haiku is about nature's creative force, and when we read the Rig Veda, we find many verses in praise of nature. Imagine one of the oldest civilisations known to humans — before the sun, the moon, and the earth had names, before there was language as we know it. Men and women must have marvelled at the colours and wonders around them. It's no surprise then that nature was worshipped in the Vedic period.

Hindus and Buddhists believe that all creation is composed of five essential elements, the *Panchabhootas*. With death, everything is transposed into these elements of nature, completing the cycles of creation and destruction.

The five elements are:

Ether — *Akasha*, in Sanskrit, is associated with *sound*.
Air — *Vayu* is associated with *sound and touch*.
Fire — *Agni* is associated with *sound, touch, and form*.
Water — *Jalam* is associated with *sound, touch, form, and flavour*.
Earth — *Prithvi* is associated with *sound, touch, form, flavour, and smell*.

This classification and the reflected thinking processes are woven into the fabric of our daily activities. These five elements are widely used in all art, including poetry, literature, dance, music, painting, and even ayurveda, the system of traditional

EDITOR'S FOREWORD

vast is the ocean of sacred words
which enlightens the universe
with divine vision
-Rig Veda

This anthology began as a *beej* in my mind — a dream project that entailed publishing in India an anthology of contemporary haiku from around the world, and making it affordable for anyone wanting to know more about haiku in my country. Haikai is blooming here and schools and colleges are beginning to show an interest in this season-based poetic form. A year's journey began when I dashed off emails to poets whose work I've always enjoyed, asking them to send me their ten best haiku. The number of poets began to swell as I reached out to poets from India and across the globe. I also dreamed of featuring the work of my own students in this anthology. I sat up in bed one night thinking, "I can't do this alone!" Sanjuktaa Asopa and Shloka Shankar graciously joined the project as editors and the amount of work they put in fills me with gratitude.

In the silences between notes, between words, between lines, the emotion that arises is known as *Rasa* — the aesthetic essence — which gives poetry, music or dance, a much greater sense of depth and resonance. It is something that cannot be described in words because it takes us to a sublime plane where even sounds have dropped off.

The 746 haiku in this volume come from 231 authors from 26 countries. They resonate, each in a unique way — just for you!

NAAD ANUNAAD
SOUND AND RESONANCE

Haiku are word paintings. In film jargon, we could refer to them as shots frozen in time. But a haiku doesn't just stop there; it pilots the reader beyond images into a sacred realm. For a poem complete in just a few words, 'resonance' becomes the keynote.

In Sanskrit, the primordial sound in the cosmos, referred to metaphorically as AUM, is known as naad, and its resonance as anunaad. According to the ancient Indian texts, an instrument, such as the voice, only resonates the cosmic sound in acts such as singing or speaking. The cavities in the body — the oral and sinus cavities and the spaces from our toes to our brain, all act as resonators or amplifiers for the 'sound' to reverberate, and this resonance is termed as *anunaad* in Sanskrit.

The poet Kabir has expressed this in one of his spiritual songs, and says that his body is a musical instrument that only transmits the timeless cosmic resonance.

wave calls …
haunting melodies linger
in the mind's abyss

$\overline{x}$

CONTENTS

NAAD ANUNAAD
SOUND AND RESONANCE   XI

EDITOR'S FOREWORD   XIII

ACKNOWLEDGEMENTS   XXIII

HAIKU   1 to 216

INDEX OF POETS AND CREDITS   217 to 235

Naad Anunaad

First Edition - September 2016

© Authors

ISBN - 978-93-85665-33-2

All rights reserved. No part of this book may be reproduced, used or stored in any form without the prior permission of the Publisher.

The views expressed in this book are those of the authors and do not necessarily represent the views of Vishwakarma Publications.

This is a work of fiction. Names, characters, businesses, places, events and incidents are either the products of the authors' imagination or used in a fictitious manner. Any resemblance to actual persons, living or dead, or to actual events is purely coincidental.

Published by:
Vishwakarma Publications
283, Budhwar Peth, Near City Post,
Pune - 411002.
Phone No: (020) 20261157 / 24448989
Email: info@vpindia.co.in / Website: www.vpindia.co.in

Cover design by Geeta Dharmarajan

Water Lilies: artwork in ink and watercolours by Alaka Yeravadekar

Calligraphy by Shipra Dutta – Likhawat

Bamboo shoots by Namrata Deshpande

Typeset and Layout by Chaitali Nachnekar - Vishwakarma Publications

Printed at Repro Knowledgecast Limited, Thane

An Anthology of Contemporary
World Haiku

KALA RAMESH
Editor-in-Chief

SANJUKTAA ASOPA & SHLOKA SHANKAR
Editors

Genuine Fake

A single *haiku* — such intimacy, a solitary consciousness in momentary resonance with a tiny detail of the universe! But anthologized, as in the wide-ranging *Naad Anunaad,* an exponential shift occurs — the encounter expands, becomes multifaceted, prismatic, kaleidoscopic; as immense as life itself.
— *Kyoto Journal*

Haiku are easily accessible poems that reveal life's meaningful moments, and more importantly, they are designed to be shared. They are the perfect antidote to technologies that threaten to consume and separate humankind. *Naad Anunaad* showcases the best contemporary haiku, connects the reader to 231 poets, and shows that the meaningful moments of a global race are not so dissimilar.
— *Modern Haiku*

I0498105

Naad Anunaad, a dream project initiated by Kala Ramesh, is the first international haiku anthology to come from India. It maps a wide range of haiku as practised in the contemporary worldscape of this 400-year-old art form from Japan. Tradition, to be alive, needs to be a flowing river, for stagnant water stinks. The so-called 'simple' haiku has been changing its face rapidly... multi-faceted and elusive, it's so difficult to keep pace with it now!

In India, as elsewhere, when we go on mountain treks, we rely on the locals to know the wind, the rain, the soil, and even how big the moon is going to be when it rises ... and if we contemplate, reflect, and explore this knowing through haiku, we may fine-tune the art of coming closer to mother nature and being attentive to her needs just as these locals have always been.

We are hoping *Naad Anunaad* will draw a new generation of readers and authors into the kind of intimacy with nature that our grandmothers enjoyed. Nature here does not mean just the hills, rivers and forests — it includes cities and the life we live, our day-to-day activities, taking in stride the agonies, failures, successes and idiosyncrasies intertwined with the natural world.

We welcome you most warmly to *Naad Anunaad* and to the world of haiku in all its various avatars.

CHAPTER 1

BECOMING A LEADER WHILE HIDING IN PLAIN SIGHT

By Anita Renee Blue

All my life, I wanted to be *someone* special. I wanted to be a part of something special. But I had a secret . . .

I excelled in sports as the Captain of my basketball, softball, and track teams. I acted as squad leader in the Junior ROTC program during my high school years. Yet, I still felt I lived in the shadows. You see, I had a secret.

I come from a small town in central Louisiana. As the eldest of four children to a single mother who worked two jobs to support us, I took on responsibilities beyond my age. I grew up regularly attending church, went to Sunday School every week, sang in the choir, and was a junior usher. Church was a HUGH part of my life. But I had a secret.

My secret was that I was a victim of molestation and I was a lesbian. Now, back in the '70s and '80s, being gay or lesbian was truly frowned upon. Growing up, I knew I was different. I had stronger feelings for girls than I did for boys. But because of the stigma, I would act as if everything was normal. My mother was a very strong woman and my HERO but me talking to her at that time was not an option.

The night of my high school graduation, one of the most dramatic events happened. I was now 18, and I wanted to be me, my true self—no more secrets. I did not go to my prom because I did not feel comfortable pretending anymore. My girlfriend and I decided to celebrate together at the local gay club. So, we are in the club dancing, happy, and having a great time when we turned around and noticed several of our classmates in the club as well. We grabbed each other's hand and ran to hide behind the curtain on the stage. Sheer panic was what I felt! We looked at each other and whispered, "what do we do?"

And in that moment of fear, I found the courage from somewhere. And in that defining moment, we decided we no longer would live in this fear. Together, we stepped from behind that curtain, hand in hand, and confirmed what they had been whispering about for years. We were together, as a couple. And in that moment of truth, I felt that I was someone special and part of something incredible.

In a word. I felt Liberated.

Life moved on. My basketball scholarship to a Junior College took me to Ellisville, Mississippi, but she stayed in Louisiana to attend college. Life and the distance would not let the relationship last, so we ended it. In my desire to do right in the eyes of God and make my mother happy, I dated a football player at college. That ended up with me getting pregnant. And I felt so out of touch with myself and reality. I went home for the summer, suffered a miscarriage, and lost the baby. It was sad but a sigh of relief because I did not want to be a statistic. You know the one: another young, teenage mother. I had placed myself in a vulnerable situation and paid the ultimate price, trying to hide my secret.

I did not go back to Mississippi. Instead, I went to New Orleans to live with my aunt and go to college there. This is where I started to come into myself. I wanted my own car, and my mother was unable to buy me one. So, I had to figure out how to get the money for myself. It was 1986, and I decided to join the Air National Guard. This started my thirty-four-year military career.

Off I went to Lackland Air Force Base for basic training. I was chosen, once again, to be a squad leader. And honestly, I was terrified. You would never know it because I was a shy person. But I persevered and made it through. What a great feeling of accomplishment. I became the first person in my family to serve. Now mind you, I had no idea I would serve in the military as my career. I could not even imagine 34 years straight. I only joined to buy my car. And I did. It was a used yellow Mazda GLC, 5-speed, hatchback. You could not tell me anything about my new car. In my mind, this was my first major goal that I had ever achieved. Zig Ziglar helped me understand my first great lesson, "What you get by achieving your goals is not as important as *what you become* by achieving your goals." I was becoming a stronger leader and did not even realize it.

My next adventure took me to Anchorage, Alaska, in the dead of winter in January 1988. Yes, it was for love, no doubt, but that is another story. But remember, I am from Louisiana, so being around and driving in snow was a whole new world for me. So, take a guess what my first job would be. A courier for the Anchorage Times, working in the advertisement department driving around town in the snow delivering ad copy. But that did not last long. I started working for Alyeska Pipeline Company in the mailroom. And by the time I left the company five years later, I was working in human resources for the benefits department as the Savings and Investment Consultant, giving out loans to the employees. Talk about moving on UP! The company provided tremendous help to me to overcome my fear of public speaking by offering Toastmaster classes during lunch. I registered for the classes to overcome that fear!

At this time, I also went through a breakup with my girlfriend, a great time of confusion about my sexuality, and in that confusion, I became pregnant once more and had a beautiful baby girl.

This entire time I still served in the Air National Guard. And I kept my secret. No one in the guard, at work, or church knew about it. You must understand, I was in the military when you could be discharged just for being a lesbian! It was later that it evolved to "don't ask, don't tell." Then, you could serve and be lesbian and even get married. So, I was forced to keep my secret. I lived every day, hiding and denying myself my real truth, for many painful years.

Yet, as life would have it, I have been blessed with many opportunities to serve in leadership positions. And what I am going to share with you now is how I grew to become a leader. But first, it is important to understand what leadership means to me.

As I mentioned, I am the oldest of four, served as a squad leader in my Junior ROTC, and played as Captain in many sports in high school. This is what I call my foundations of leadership. I was young and already exhibiting leadership characteristics. With each leadership position, I have risen to the occasion and grown because of the responsibilities demanded of me in that role.

"The most dangerous leadership myth is that leaders are born - that there is a genetic factor to leadership. This myth asserts that people simply either have certain charismatic qualities or not. That's nonsense; in fact, the opposite is true. Leaders are made rather than born."—Warren G. Bennis

Leadership is the art of flexibility. It is being able to adjust and communicate in different ways, specific to each person within your sphere. You must exhibit enough self-awareness to know what is going on around you and, at the same time, yield the best response from each person. This demands patience, empathy, and compassion.

And now, with thirty-four-years in the military as the basis of my experience, the following characteristics are what I feel to be the critical qualities of acting as a good leader:

Commitment
Clarity
Courage
Passion
Humility

The ability to act decisively reflects **Commitment**. When serving the Guard in Alaska, I was promoted to the rank of Staff Sergeant or E5. I was responsible for the storeroom in the dining facility. Now before this promotion, I wasn't the best Airman. I was late for duty and often called in sick or was simply not in attendance. But once I was placed in this leadership role, I knew confidence had been entrusted in me. And the feeling of that responsibility demanded that I had to be present and on time. I had to step up! I had to commit to becoming that better version of myself. A leader! And this evolution started with this first promotion. Thinking back on this key event, I recall what Ralph Ellison wrote, *"It takes a deep commitment to change and even deeper commitment to grow."*

When a leader has **Clarity**, it allows others to digest their goals and decide whether they will support the leader's cause. In the leadership role as a military member, I was trained to give detailed instructions so that my subordinates were clear about my expectations. And this clarity then assisted them in growing into their leadership roles. I gave and received feedback just for that purpose. It has been my experience that very few people know what they want, much less know how to get there. So, they gravitate to those who have vision, a clear picture that can be shared in the minds of those around them. Jim Rohn wrote that clarity leads to great achievement. *"Take advantage of every opportunity to practice your communication skills so that when important occasions arise, you will have the gift, the style, the sharpness, the clarity, and the emotions to affect other people."*

Demonstrating **Courage** as a leader is a BIG key. No, it is more than that. Demonstrating courage is absolutely critical. I have been in the military in three different states: Louisiana, Alaska, and Texas. I started out as a very shy person in life. But I had to become bold and practice acting courageously to be courageous. Moving through the ranks while facing adverse risk as a black, lesbian female meant always hiding in thought. It meant not being able to truly express myself. It meant living with a secret that could instantly end my career. And in the end, this courage helped me to fulfill my role as a good leader. You must remind yourself that your time is limited. You cannot waste a moment living with the results of other people's thinking. Refuse to allow the noise of the opinion of others to drown out your inner voice. And like Steve Jobs reminds us, *"have the courage to follow your heart and intuition."*

My **Passion** has always been shown in my love for people. How they are treated is rooted in my own experience of having to keep my secret and never truly having a voice. For thirteen years, I served as an Equal Opportunity Craftsman in the military. There I was able to train on Diversity, Inclusion, and Cultural Awareness. This role meant fighting for people in areas of sexual harassment as well as discrimination based on race, color, sex, gender, and national origin. And during my training for this career, I was sexually harassed. And I had to report it. He harassed the wrong person. And after a long fight, sexual orientation has been added to the protected classes for military members. And this gives me *hope* like the words of Maya Angelou, *"My mission in life is not merely to survive, but to thrive; and to do so with some passion, some compassion, some humor, and some style."*

And last but certainly not the least, I feel that to be a good leader, **Humility** is a MUST. It can be hard to admit your weakness or vulnerability, but a good leader will demonstrate it when required. Confidence is an attractive trait. And there is simply nothing like a humble character for creating a lovable persona. A good leader can *admit when they are wrong* and *take criticism as an opportunity for growth*. Michelle Obama's perspective inspires me every day, *"We learned about gratitude and humility - that so many people had a hand in our success, from the teachers who inspired us to the janitors who kept our school clean . . . and we were taught to value everyone's contribution and treat everyone with respect."*

Each of these characteristics have brought me to where I am today. And they will continue to carry me forward on my path to being one of the most outstanding leaders. And as a great leader, I aspire to become one of the greatest entrepreneurs in history. Leaders engage. They try different paths, like many of the business opportunities that have allowed me to grow tremendously. Each was like a stepping stone, making me stronger for the next one.

Knowing that I did my best in my mind just does not work for me. Instead, I look at each event as a chance to grow and pay attention to timing. Hindsight is 20/20, they say. And I believe that to be true in my case. You must stay the course! And whatever you try to accomplish, even if it does not turn out as you expect it, remember that we are all destined for something great in our lives. We just have to believe, keep the faith in ourselves, and fight through each challenge to find our own meaning of Victory and Success.

BIOGRAPHY

Anita Renee Blue is a thirty-four-year air force veteran, mother, wife, businesswoman, and an active member of her community. She has held many leadership positions in the military and as a civilian, adding to her extensive life experience in this arena. She desires to continue to be a servant leader to people throughout the world. She is highly motivated by seeing others achieving their goals and dreams through their commitment, courage, clarity, passion, and humility. Challenges will come, but she believes that whoever stays the course will eventually win the race of this thing called Life, with love in their hearts and joy in their souls.

Anita Renee Blue can be contacted via https://linktr.ee/AnitaBlue

CHAPTER 2

How To Deal With An Unsupportive Spouse In Business

By Matt Morris

In the process of leading hundreds of thousands of entrepreneurs and salespeople over my twenty-five-year career, I've witnessed that relationship turmoil is one of the single biggest killers to success in business.

Nothing saps your energy or drains your spirit more than disharmony in the household.

Before we dive in, I want to make it clear . . . I'm no relationship Goo-Roo.

I'm divorced, so taking relationship advice from me might be a bit like learning how to get six-pack abs from a sumo wrestler.

So, while I'm no relationship expert, I am an expert in human nature. I've coached thousands of entrepreneurs over my career, and relationship disharmony that impacts negatively on business development is one situation that I've been able to coach many others through successfully.

Many of you reading this may not have this issue or may not even be in a relationship, but you must read this anyway, because, on your journey to leadership, this WILL undoubtedly be an issue for some of the people you lead.

My training here is assuming you're in a committed relationship, either married or headed in that direction.

7 WAYS TO OVERCOME AN UNSUPPORTIVE SPOUSE IN BUSINESS:

#1 Don't push

Are you pressuring and pushing your spouse to be as excited about your business as you are?

I know your first instinct is to say, "No, of course not."

But, if I were to ask your spouse if they're feeling pressured, what would they say?

If you suspect there is even a slight hint of this, I'd recommend you ask them and get their perspective on things. If you want them to understand your dreams and desires in business, then I suggest you follow the sage wisdom of Dr. Stephen Covey: *"Seek first to understand, then to be understood."*

The more you push, the more likely it is that you'll create conflict. Some people aren't wired for entrepreneurship or, at least, don't believe they are because of their beliefs. Instead of pressuring them to be excited, you should be you and let them be them.

If you want your spouse to support your dreams and desires, you first must try to understand their dreams and desires and make sure you support them, even though it might be difficult at times to fully understand each other's goals.

#2 If you want support, give support

How exactly are you giving support in the ways that your spouse wants the most?

Notice, I said, "in the ways that your spouse wants."

You may be providing money, a home, doing the cooking, and cleaning. Those may be important for you, but, in many cases, spouses aren't supportive because they don't feel supported in the ways that matter most to them.

How can you BE the example for support that you want to be? The answer may not be obvious to you, so you may need to dig in, ask, and work on resolving that question until you find the answer.

To sum it up, you must be the example.

#3 Neutralize the threat

In many cases, an unsupportive spouse is unsupportive because they feel as though your business is a threat to the relationship. They may feel that you consider your business a bigger priority than the relationship and that your business is pulling you away from them.

You need to understand that this threat may not be vocalized because your spouse doesn't want to appear insecure or controlling. However, the reason they are not supportive could be because of their underlying fear and insecurity about the relationship.

For many entrepreneurs, this is hard to understand: in your mind, you're "doing the business for them" or for the family, and you've told them that. However, you must realize that even though that might be true for you, and even though you've told them that time and time again, they aren't feeling it.

So, how do you get them to feel it? Keep reading . . .

#4 Honor them

Find ways to admire and acknowledge what you honor and respect about your spouse.

Whether we want to admit it or not, we all have a desire to be respected. If your spouse feels as though you're respecting your business or your business partners more than them, it will drive a wedge between the two of you.

If you want to be supported and honored as an entrepreneur, support and honor your spouse and other family members from whom you want support.

#5 Appreciate them.

On a daily, yes, DAILY basis, take steps to ensure that your spouse feels genuinely appreciated. Of course, it's great to tell them, but folks generally see better than they hear, so make it a regular practice to write little letters or notes of appreciation for them. Even if it's just a sticky note, it shows that you care and make them feel appreciated and loved.

When you're out working on your business, take a few minutes to send them a text message appreciating and acknowledging them. I've seen relationships transform through the simple act of an appreciative text or sticky note sent every single day.

If you feel as though you've tried everything, then try writing a text or a note to send them every single day for the next thirty days and see what happens.

Reward the behaviors you want.

When your spouse shows any sign of support, go overboard in showing your appreciation for that support.

#6 Unconditional love

In our heart of hearts, what we all want is unconditional love—to be loved no matter what.

When your spouse is unsupportive, understand that it's likely because of their fears, their lack of feeling supported themselves, their lack of feeling honored, appreciated, or loved. So, show your love for them, no matter what.

If you want them to support you, whether you're winning or not, love them whether they're demonstrating love for you or not.

Act to ensure that your spouse understands that they are the most important thing in your life and show that you love them, no matter what.

#7 SUCCEED

I've seen hundreds of people go from unsupportive to being their spouses' biggest fans when they see their spouses winning.

There's an old story from Tom "Big Al" Schreiter that fits here . . .

A husband gets involved in network marketing. He's super-excited and has a dream of buying a new Cadillac. He cuts out a picture of a Cadillac and tapes it to their bathroom mirror. But when he gets home that night from work and a busy day of doing meetings, his unsupportive wife takes the picture down and puts it in the drawer.

But, undeterred, the husband tapes it back up on the mirror. Then, when he comes home the next night, again, it's in the drawer.

This process repeats every day for months.

Until one day, the husband comes home with a brand-new Cadillac, bought with money earned from the profits of his new business.

The next day, when the husband comes home, taped to their bathroom mirror is a picture of a mink coat!

Your spouse wants to see you win, but they might not believe it's possible.

So, do all the things necessary to win.

BIOGRAPHY

Author of the international bestseller, *The Unemployed Millionaire*, Matt Morris began his career as a serial entrepreneur aged eighteen. Since then, he has generated over $1.5 billion through his sales organizations, with a total of over one million customers worldwide. As a self-made millionaire and one of the top internet and network marketing experts, he's been featured on international radio and television and spoken from platforms to audiences in over twenty-five countries around the world. And now, as the founder of Success Publishing, he co-authors with leading experts from every walk of life.

Contact Information
Website: http://www.MattMorris.com
Company website: http://successpublishing.com/

CHAPTER 3

THE PERSON YOU COULD HAVE BEEN

By Steve Moreland

If Fate's blood-stained cauldron has not found your life yet, she's hiding just over the horizon, waiting until you're at your most vulnerable. So if you're willing to listen to someone that knows about life's ash heap, I'll share the Lessons I learned after I failed my Test. The lessons focus on our thinking. More specifically, about how thinking differently empowered me to thrive where most cannot imagine surviving. I promise not to waste your time with fluffy bullshit or rah-rah! Just the mental tools what worked, that brought me across a desert wilderness of 5,544 days.

May the following battle-tested advice return you from your seemingly impossible cauldron *"tested—and found not wanting."*

We Texans pride ourselves on our Code. Toughness is Rule #1. And it means *"no tears allowed."* See, our cult-like indoctrination begins the moment we are born. And the other Spartan rules include: *do only BIG things*, especially if others say it can't be done; *rub some dirt on it* because blood and scars prove your worth; and *do Right*, even if the Lord God, himself, threatens you to do otherwise!

Brutal. Absolutely! But definitely the kind of folks you'd want covering your back in a fight. It's a belief carved deep in our soul—that there simply is NO FREE LUNCH. It is a creed rooted in commitment and perseverance, summed up in one word. Grit!

The standard we have to carry begins early. At age twelve, I started *"earning my worth."* My phone rang off the wall with grass-cutting jobs in the Texas infernos called summer because my dad drilled me to do what everyone else is afraid of, to deliver results beyond expectations. Just self-disciplined results! No excuses.

I went right to corporate America after graduating with academic scholarships – working for three Fortune 500 companies before I was 24. At twenty-five I was in charge of my own brokerage firm in Dallas. By thirty, I'd made it to millionaire status, flew in private jets, brokered 9-figure deals from European castles, banked in numbered Swiss accounts, and spoke on international stages raising millions for venture capital deals.

Ballistic was my term for the vertical climb I experienced. Simultaneously serving as vice president of offshore operations for a boutique hedge fund, CEO of a 58-office tax and wealth management firm, and co-principal of a SaaS startup. I couldn't afford the luxury of sleep. And part of every month, I lived near my office in the banking district of Nassau, Bahamas, acting as the vice president of business development for a middle eastern banking syndicate.

Occasionally, I woke up at a place my then-wife and children called home. It was there that I slowed down enough to rub some of that Texas dirt on my hand tremors from sleeping only on those overseas flights. I was stumbling forward just to maintain the pace.

There was something wrong but I could not risk failing the mission. My Dad's standing orders were crystal: *"You can rest when you're dead!"* And this belief came from his creed that a man only earns a medal on his gravestone if he dies "in combat."

Well, I failed to become a "lifer" in the Corps, so I determined that I was going to achieve whatever most would call impossible. I believed in his invincibility! And after eighteen years of his Marine-style bootcamp, I feared only one thing, **"meeting the person I could have been!"**

So, when Fate's blood-stained hurricane came for me, I was Ready. Ready to blindly march into Hell itself. But after the first few years, I felt more like the Greek myth of Sisyphus who was sentenced to pushing a boulder up the mountain every damn day and then waking up the next morning to find it at the bottom again. I remember thinking to myself, "Maybe God is *not* good" after feeling soul-crushing agony for the first time. Real pain that made me wish I could just die and get it over.

I'll admit, all that invincibility crap did NOT work. And I'm painfully embarrassed to admit that I found myself wallowing in my self-pity after losing absolutely everything and feeling abandonment by all I loved. I had succumbed to that state of a *victim*. And you know what, that Texas dirt did NOT fix the wounds I'd caused my family for the undeserved trials and tribulations my bull-headed foolishness caused.

Though I was brought up with my dad's relentless Texan and Marine Corp code of conduct mixed with my mom's Christian beliefs, the devasting pain caused me to question their beliefs. Sitting in the ash heap of my life like the Bible's character Job, I commenced to blaming God for not protecting us from the horror that imprisoned us. I begged and even prayed for an instant release of misery, even raising my fist in anger and shouting "You're *NOT* a good god!"

I just wanted that magical snap of a finger and everything to be like it used to be. But genie-like fixes never happen, do they. Why? Because strength is *not* forged in luxury and comfort. Medals do not get pinned to your chest for holding hands and singing "Kum Ba Yah."

The struggle to endure real agony, to eat suffering, and know your pain so intimately that you name her has a purpose. You see, it took time for me to get over my self-entitlement in order to face my demons and do the most excruciating thing I'd ever done. Realizing that I could not change the past or erase what my mistakes had cost my family, I had to make a decision: either continue to blame others and wallow in self-pity or use the hell I was inside to forge a better version of me!

In school, we're first taught the lesson that prepares us for the test. But, in life, we face the Test first; later, we learn the Lesson.

The grade is what we become through it all. It's pass or fail. And yes, hell is when you meet that person you could have been. It means rising again and again within the blood-stained cauldron of Fate. Only this repeated discipline distinguishes the few from the many, the extraordinary from the ordinary. The worthy from the worthless.

But that person you could have been is only Hell if he or she stands better than you chose to become! **Hell, then, is meeting the *better* person you could have been.**

Like I promised in the beginning, what follows WILL take you through any hell. And you will arrive on the other side, *"tested – and found not wanting."*

Let's begin with a question: "Have you ever been really curious about something—to the point of obsession?"

Since I was a kid, I wanted to unravel this thing called thinking. I reasoned to myself that if I could only understand how the few we call "great" actually thought, I might be able to be like them and make the world a little bit better. Because, for the most part, they are human just like me. The only difference is that they *see things differently* in their minds.

Personal development "coaches" blather about managing our thinking. It is THE key, agreed. But it's not enough to know *what* to do. We've got to know *how* to do it. It's the subtle and often hidden difference between learning science without the art of knowing how it applies to real-world situations. Most of these "well-meaning" coaches deserve an "A" for science but an "F" in art. Never earning a medal from within Fate's blood-stained cauldron means their theories can get you to one destination – that chance to meet the person you could have been.

Here's an example of a coach with earned rank, Dr. Viktor Frankl – author of *Man's Search For Meaning*. Frankl didn't just survive six years of Nazi concentration camps, he changed the world forever with his discovery of how we create meaning through our imagination.

Better thinking creates better doing. And better doing creates a better being.

Frankl forced me to think. I mean really think. And all of a sudden, what Professor Eli Goldratt wrote in *The Goal* became crystal clear. "If we continue to do what we have done, which is what everybody else is doing, we will continue to get the same *unsatisfactory* result." But I asked myself, isn't that what we do so very often - more of what everyone else has done and then expecting a different outcome?

We are what we've done, right? So, aren't our own actions - what we *do* - that creates who we *become*? In short, "doing creates being." So, who we are today – our being, is a product of our past doings? Becoming someone better can only happen by doing differently. And differently results from the seed of the thoughts in our imagination.

Because I wanted a different future – one that honored the sacred by making the world better, I could no longer afford to think like I used to, or like everyone else. Maybe you're brighter than me and already know this. But for me, this realization was the Eureka! And in that realization, I felt something deep inside like lightning.

If my prior thinking caused my current doings (my actions and habits that are known as my reality), **then why couldn't I change my future by changing the way I was thinking?**

Socrates (Greek philosopher 470 B.C.) taught a Secret passed through his student Plato to his student Aristotle (Greek philosopher 384 B.C.). Aristotle planted this Secret into the mind of a 13-year-old prince. This Secret method of thinking changed the ancient world.

At 16 years of age, the prince led his cavalry at the Battle of Chaeronea, decimating a supposedly unbeatable enemy. At 20 years of age, he became king of Greece, marched his army towards Persia, solved the riddle of the Gordian Knot, and destroyed any that opposed.

At 24, he captured the supposedly unconquerable city of Tyre. At 25, he became Pharaoh of Egypt and then returned to the desert near modern-day

Babylon to lead his 50,000-man army against a force exceeding 500,000 led by the Persian emperor Darius. Charging into the front line on his legendary black stallion Bucephalus, he achieved the impossible and became emperor of the known world.

By age 30, he had created the largest empire in history. Today, he's still studied in war colleges for his battlefield genius, ethical governance, and unrivaled valor.

The Secret thought? "Be as you wish to seem."

The Result? One *impossible* difficulty after another - CRUSHED!

His Name? Alexander

How is he remembered? Alexander—the Great!

In school, we're first taught the lesson that prepares us for the test. But, in life, we face the Test first; later, we learn the Lesson.

Here's my experience. The Lessons learned *after* the Test lead to better actions—which lead to becoming a better being, right? That means that tests uncover our weaknesses so that we can learn greater lessons. What and who we become through the Tests reflects our grade in life.

If we're honest, we'll admit that we often create our own storms. And then we blame others when we must endure them. But if we use the agony, we find something called grit. Grit is commitment bathed in love to become better than we were the day before. It's a relentless dedication to rise—to become better, stronger, and wiser. It's a refusal to quit, even when we feel we can't get up again.

The question is, will we? Will we persist after the problems that were caused by our poor thinking – and the results that followed? Or will we just quit due to the fear of failing and the probability that life won't be easy?

Being *"tested and found not wanting"* means we'll certainly be scarred from one battle after another. But the scars reflect rank, defining how many times we returned to the cauldron instead of hiding and waiting to be rescued by the God that's testing us.

It may be cliché, but our very thinking sparks our every action. Put another way, our doings, added together over time, construct our being - *what* and *who* we become.

Do we dishonor the Sacred, settling for what everybody else is doing and continuing to get their same *unsatisfactory* results?

Or do we ***think* better**, in order to ***do* better**, so that we could ***be* better**?

We become what we choose to be. This is the Secret. My gift to you, as Aristotle long ago shared with Alexander, "be[come] as you wish to seem."

Now you know that Hell is NOT meeting the person you could have been.

Hell is meeting the *better* person you could have been.

BIOGRAPHY

A native Texan, Steve Moreland is known for two things. Dedicated practice. And success. Success equates to one's level of practice. So he really does only one thing. His Rubicon system teaches how to perform the common under uncommon conditions.

Motivated by the Latin creed FORTES FORTUNA ADIUVAT – "Fortune favors the brave," his mission is to deliberately cause affirmative outcomes that would not have occurred otherwise.

Connect with Steve via LinkTree: https://linktr.ee/steve_moreland

CHAPTER 4

THE MAGIC OF TIME

By Arjan Scholten

Time is life's great enigma. What is more mysterious than time? But what exactly is it, and what does it mean to us, as humans? And, perhaps most importantly, how does it work? In the course of my life, I've spent a lot of time thinking about time—and I believe that many of us are driven from time to time to ponder about the puzzle that time, as we know it, presents to us. And now, I'd like to invite you to spend a little time with me.

I was born in the Netherlands, and, as a young boy, suffered from a disease called asthma. It was challenging for a child to live with the disease. Although, in time, the disease would prove to be both a burden and a blessing in my life. When I was in my first year at primary school, I spent more time either at home or in the hospital because of the asthma. That was also the year that my parents got divorced. Because of the illness, I was made to repeat that first year of school, to make up for the time I had lost. You can imagine how I felt; it was a tough time, and it was difficult to connect with my classmates. I remember thinking that doing the same year over again at school meant that I would lose a year of my life. It's generally true that when you're young, you want to grow up fast. You want to go to bed late and do all the things that grown-ups are doing. Time never seems to go fast enough for you. Or at least, that's how it feels.

Over the next few years, my asthma became a real issue. I was 10, and my parents decided that it really needed to be dealt with. So, they started to search for a healthier environment for me to live in, hoping that things would improve and I would learn to live with my asthma. They came up with two options: One was an asthma center based in Switzerland, the second was another clinic located in the Netherlands. That's how I ended up staying for 15 months in the clinic in the Netherlands, far away from my parents, family, friends, or anyone familiar. I felt lost and often wondered how the time spent in the clinic would benefit me. But, at the same time, I began to realize that maybe I should look at the experience differently. Like it or not, the one thing that never changes for all of us, human beings, is the fact that you live by yourself and, therefore, experience life by yourself. This was an incredible insight for me. At that young age, I learned that I alone was responsible for connecting with people, forming my opinions, and living my life in a way that made sense to me. Some people might call this a kind of freedom. It was undoubtedly true that the idea of the constraints of time took on a completely different meaning for me than previously. I realized that there is no such thing as a time constraint because it's only a matter of what you believe. How you spend your time and what you do during the time you spend in this world is entirely up to you. Based on my experience, my advice is to spend it wisely, for time is not coming back.

And so that is what I tried to do. I became the master of my own plan, not an organized plan, but one that unfolded based on life experiences. I finished attending MAVO secondary school in the Netherlands at the age of 17, and wondered: What do I do now? Again, time became an issue for me. It was expected that, like most kids, I would find a course of study, go to college, and eventually get a master's degree in some field. But was that really what I wanted for myself? I mean, what did I know of life? It goes by so fast and, besides, I had questions about what kind of life choices I should make. It seemed crazy to be expected to make choices about the right educational path to take. I thought that it might be a good idea for me to take some time to find out who I was as a person.

But could I actually do that? My mind was full of what-ifs. You probably know that feeling when your emotions seem to take over, and your good judgment becomes clouded. I felt so strongly that I needed more time to get to know me. So, I decided to leave my comfort zone and seek adventure! I still think it's really important to find out who you are and how you tick, and only then make the big decisions that will affect the rest of your life. So, at 17, I decided that, since life was happening all around me, I should do what I believed was going to be good for me!

I chose to find myself, to get familiar with the real Arjan and find out what he's about, before making any of those big choices. Up to that point, my life had been dictated by my upbringing, my disease, and a somewhat limited view of the world. But my thoughts eventually led me to take a great leap forward into the unknown and embark on a big adventure: I decided to go abroad and live for a year in the United States, where I would attend high school. That would give me a chance to see how well I could manage and learn more about myself. It was a scary thought but, at the same time, full of promise and with the potential to be very rewarding. You could say that I took a whole year out to study myself, Arjan Scholten. What a great decision that was! By taking that time out, I learned so much about life and myself. I'd say that it saved me years of struggling to find my purpose in life, and gave me time to shape my own beliefs and goals that I could stand by. My life plan began to emerge—or, at least, I felt that I had an idea that, if acted upon, could become a life plan.

At last, it seemed that my mission in life was clear. Best of all was the realization that I had the whole of my life to make it happen. What a gift, and what a time saver! Unlike many other people, who go through life, not knowing what to do, when they're older, are left saying: "God, I wish I'd done so-and-so, but now I'm too old, and there's no time left to live my dreams." Not for me!

After spending a year in the US, I returned to the Netherlands, a reborn young man of 18. During my time away, I'd discovered that I had skills and decided that I would use education to help me master them. I went to college and eventually earned myself a bachelor's degree in commerce.

I started working and became a CRM expert—that's Customer Relations Management. I believe that everything in life, whether it's to do with business or your personal life, is about the dialogue we have with one another. We, humans, communicate with each other all the time. That's how we relate to people—by giving time to each other. Back then, I felt strong and worked hard on the goals I'd set for myself, but, unfortunately, I forgot my health and the fact that I have asthma. I was working hard and playing hard and generally enjoying life. But, while doing that, I forgot that my system, my body, needed proper care and that my life-work balance needed to be in sync. I needed to establish healthier eating habits and exercise more often. Things came to a standstill when I was 32 and was hospitalized with severe lung problems. Before I knew it, I found myself in intensive care, multiple tubes going in and out of my body, and fighting to survive.

The doctor told me, when I came in, the condition of my lungs was like that of an 80-year-old. In other words, like those of someone who had not taken their health seriously at all.

I am very fortunate to be alive today. After spending three weeks in the hospital, I was finally released and able to go home. But, what a time I had at the hospital! When I was conscious and transferred from the intensive care unit to normal medical care, I had a lot of time to think. I lay awake for hours in bed, my life flashing before my eyes, as I realized that this might be it—my life could be over. I could have died right there and then: A young man who died at 32 due to a severe asthma attack. What a way to go! That was also the moment that I finally understood the wisdom of "health before wealth."

That was a hard lesson for me to learn and comprehend. It took me another six months of being at home to come to my senses. My pride became anger, then fear, and finally gratitude, gratitude for life, for just simply being around and, hopefully, living happily ever after. So, once again, I made a life plan.

That plan contained everything. I'd learned from my experiences and the reason why I made the decision to enjoy life, to make the best of the time I have. Now, I make it my mission to inspire and help people by giving them—time:

- Time as a gift.
- Time to listen.
- Time to explain.
- Time to learn.
- Time to experience life.
- Time to feel.
- Time to express themselves.
- Time to love.

That's what I have learned: Time is magic. Scientifically speaking, you can say that a day has 24 hours, a week has 168 hours, and a year lasts for 365 days, which accounts for 8,768 hours. The United Nations estimate of the global average life expectancy in 2019 was 72.6 years; if we round it up—73 x 8,768—that equals 641 thousand hours. What do you mean, you don't have time? You have all the time in the world! But that isn't all: Our experience of time is governed by our perception of time. As we grow up, our view of the world and, thus, our perception of time changes. Isn't it a true blessing and gift that time is so amazing? And it can be spent in any way we choose. But just bear in mind that once spent, that time won't come back—so my advice is to spend it wisely.

We forget that time is just an instrument, and that our lives and our time occupy only a short period from a universal point of view. We humans feel, love, and experience all sorts of other emotions. And that is a good thing. That makes us special. So, be proud of yourself, and don't worry about the choices you've made or are going to make. Time is experience, so experience it. I believe that everyone has their own life journey and that the most important thing is to make it count.

Of course, people live and die, so why not make life a great journey? Take time to think about what you want your life to be about. How would you like to be remembered? What will be your legacy? Remember, your life is over before you know it. But, as long you're here, enjoy it and make the best of the time you've been given. I wish you all the best. May this chapter serve you as time well spent.

BIOGRAPHY

Arjan Scholten is a CRM expert whose passion is to help people and businesses, educating them about how processes can work in their favor. He's a big fan of personal development and learning how to have more fun in the workplace, and life in general. His mission is to inspire people and help them live life to their full potential. He believes that, no matter what your past, you can move forward and live your dream, simply by working passionately towards achieving it. Arjan's motto is: Live your life and make it happen.

Contact Information
Facebook: https://www.facebook.com/ArjanScholtenTime
Instagram: https://www.instagram.com/arjanscholten71/

CHAPTER 5

GOING AGAINST THE GRAIN

By Ben Dahl

I still remember looking at that number, thinking, "I can't afford this." The $2,000 was more than I had in my bank account, and it would be the first time I had spent that much on anything. There I was, pacing around my disheveled room in a run-down college house in my torn sweatpants. It was an environment best suited for living life between beer cans and bottles of dip spit, but I was somewhere else in my mind. I believed I could do something more, and I was about to make one of the biggest decisions of my life.

That is not to say that things were bad. I had a part-time job at a digital marketing agency in downtown Cincinnati. My boss had practically guaranteed me a position after graduation if I stuck around. I had earned membership in the most prestigious (and fun) fraternity on campus. I was dating an intelligent, kind, and beautiful girl. My social circle was so big I couldn't leave the house without seeing one of my friends. I rocked six-pack abs, held a position in the best business clubs on campus, maintained a 3.9 GPA in my honors program, competed in intramurals, and was hitting my stride in guitar practice.

Forgive me for sounding arrogant—life was good. You're probably thinking: "Dude, you had it made!" and I did, in a way. I had it made insofar as I could have easily gone with the flow and achieved a relatively standard, white-collar quality of life without having to try.

So, what on God's green earth could possibly lead me to take all of my money, including student loans, to buy a $2,000 online course on how to start a consulting business?

The simple answer is that "I wanted more."

Like so many of us, the path I was on at the time didn't reflect my dreams.

Do you ever feel that you were destined to do something great, something the world would recognize, something that would make a difference?

I think this feeling is familiar to you, to me, to the greeter at Walmart, and to the cashier at McDonald's. I believe that we all desire greatness and fulfillment in our cores of being, even if we don't know what that means for us yet.

It's a trope, but we all have our versions of success. I didn't know what mine was at the time, but I did know was that despite the positive momentum in my life, I couldn't settle for less than I was capable of achieving. I had heard tons of stories about kids my age and younger earning themselves hundreds of thousands of dollars from their businesses, and they had all of the freedom in the world to mold their lives to their liking—I wanted in.

So, I did it. I must have walked a mile in that little house before I pulled the trigger, but I did it—I bought the course.

Betting big on yourself is a strange feeling. It's somehow easier to bet on a football game, a horse, or even a poker hand. When you bet on yourself, it's do-or-die. Everything is in your control, and success or failure comes entirely by your own doing. There is an immense feeling of freedom mixed with a dose of "what have I done?" but whether you find yourself confident or afraid, there is only one course of action—to move forward.

I dove into the course, waking up early and staying up late to take notes and learn the material. The first item of action was to create a vision and a plan for execution. The following weekend, I skipped the big fraternity party and locked myself in the library for two and a half days to avoid distraction. I must have seemed like such a loser, but that was the weekend my world changed.

That weekend, I hardly checked my phone. I visualized my dream life and put it into a PDF that I printed and kept by my bedside. Once my dream life had been mapped out, I created a one-year plan to keep me on the right path, with detailed instructions on what to do each day for the next 365 days. In theory, all I had to do was follow my own instructions to get what I wanted.

My plans changed about two weeks later when my boss sent me to a business marketing meetup to hear an expert speak on new marketing strategies. After some tertiary research, I discovered that the speaker had spent one BILLION dollars in online ads. He was also running ads for the Golden State Warriors in their prime, and he had worked with dozens of other Fortune 500 companies. I prepped some questions, sat in the front, and took notes. That presentation shattered my paradigm of how the Internet worked.

After the presentation, I stuck around to say thank you and introduce myself. I shook the lecturer's hand with all of the eager energy I could muster and pitched my vision to him. It must have made an impression on him because he told me to apply for an internship, and I got the position.

I still remember the morning I went to work to tell my boss that I had to hit the road, effectively forfeiting my guaranteed job after college. It was tough, but I had to trust my gut and seize the opportunity.

Things accelerated from there. Within two weeks of meeting my mentor, I was flying around the country, engaging with some of the biggest names in marketing on the planet, and working on international brands.

I flew out to California for a marketing conference where I met the organizer. He assigned me the position of lead marketing specialist for a professional sports team and hired an international team under me. We were even featured as experts in training videos at GoDaddy's offices in Scottsdale. My mentor personally coached me, day in and day out, and I was learning faster than I ever had before.

The coming months tested me in every possible way. One night, while visiting family back home, my brother and I were walking the dogs around 10:00 p.m. when my phone rang. It was my mentor on the line. His plane had turned around on the way to a conference, and he needed me to fill in for him the next morning in New York City. I had never presented the material before, and my mentor was a 25-year industry veteran. How could I possibly fill his shoes? Nevertheless, I rushed home, booked my flight, practiced until 4:00 a.m., and woke up to catch my plane at 5:00 a.m. I made it to the conference to deliver the presentation with about an hour to spare.

One time, I worked two back-to-back, 95-hour weeks in front of the computer, not counting bathroom breaks. Moments like these pushed me to rise to the occasion. I had to learn on the fly with limited time and focus and roll with the punches. I also learned the power of a good mentor. For the first time, I saw the life lessons I had learned. These experiences must have shaved a good decade off my traditional learning curve—in business years, I would be pushing 40 soon.

I took a semester off school to focus on work, but in conspiring fashion, things came to a halt. I worked excessively, and my relationship floundered because of it. My mentor had taught me so much, but I couldn't keep up, and I stepped away. I did too little too late to save my relationship, my lease at the college house was up, and I moved home without a job, no girlfriend, and a semester behind in school. To save money, I also stepped back from my fraternity, which would cost me the friendships I'd spent so much time building. What had I *done*?

I tried to start a new business with some guys I had met on the road. To avoid the shame of returning to school as a failure, I took a semester of online classes from home at the same time. It wasn't long before the business ended in arguments, and I was forced to return to campus or lose another semester. Those were dark months, and my new college house seemed to reflect the state of my spirit. I moved in with roommates who left food rotting on the ground next to the trash can. Talk about high-class living! I stayed up late, distracted by TV shows, and I even picked up smoking to affirm my internal dismay.

It was in those months that my bank account went all the way down to $0 (actually less than that because I overdrew my debit card), and I had to ask my parents to cover my basic living expenses for a month. That hurt. It wasn't long before that I was networking and meeting some of the most successful people in my industry. How had I sunk so low?

There were moments when I wanted to throw in the towel, finish my degree, and "get back on track." I even wrote a modified vision document that stated I would "finish my degree no matter what." I stopped trusting myself.

I couldn't bring myself to get a job, however. I couldn't "unsee" the success I once had. I couldn't silence the burning question in my mind: "Why not me?" It turns out that a yearning spirit is not easily crushed. Instead of giving up, I picked myself up, started to network, and managed to secure some projects that helped pay my bills.

Then, I got a surprising message from my mentor, offering me a second chance. How could I say no?

I finished my semester at school and went back to work with my mentor. It was just as challenging as before, but I was prepared. We flew around the world and back, working on massive projects. On one trip, I expected to be in Vegas for a two-day conference but ended up staying away from home for two months. Now, *that* is rolling with the punches. I went coast-to-coast across the United States, overseas to Taiwan, and even as far as the Philippines. There were times I slept on airport floors between 18-hour, non-stop workdays, but it was fun. My mentor helped me to save several more years on my learning curve.

The time eventually came for me to leave the roost and go out on my own with my newly found confidence and stronger skillset.

It wasn't a direct path to true entrepreneurship, however. I worked temporarily with another marketing agency to get back on my feet, though my first few personal clients were so challenging that I almost threw in the towel again.

I thought that maybe if I got a job for some security, I could grow my business part-time until it was "safe" to go out on my own, so I applied for a job and got an interview right away. An hour later, I had talked my way through three levels of management to the director of marketing at a $275 million company. They offered me the job with an annual salary of $50,000 per year with benefits, a paid vacation, flexible hours, and a "great opportunity" to earn a whopping 5% yearly raise. Now, I'm no mathematician, but 5% of $50,000 is only a couple hundred bucks extra per month *before* tax. That was when I knew that I could never work for anyone else. I knew that, if I focused my efforts, I could earn a 5% raise every day working for myself.

That was the last straw. I'd officially given up everything I once had—my degree, my girlfriend, my fraternity, my college friends, and my career—but something interesting happens when you abandon one path for another. At each major juncture along the way, you learn something new about yourself. Each of these lessons helped me mold and clarify what I want out of life, and I got closer to my vision with each step.

I met another girl who is beautiful, intelligent, and kind. I realized that no one cared if I had a degree. I learned not to work for money, but to make money work for me. I have a better appreciation for the friends I've kept. I'm in a fortunate position in that I have the ability to select the projects on which I want to work and the people with whom I want to work.

It's easy to follow the crowd. It's easy to do "what you're supposed to." It's easy to dream about having a better life. It's hard to bet on yourself. It's hard to stick it out and do the work. It's hard to pick yourself back up and break free of the chains of fear and doubt.

I knew I would never realize my dreams if I followed a traditional path. It was only by going against the grain have I given myself the opportunities to turn my dreams into reality.

BIOGRAPHY

Ben Dahl is an author, speaker, and business coach who has helped Fortune 500 companies grow online. Having spent millions of dollars in marketing and leading international teams, Ben has the technical and leadership experience to produce massive results. With his expertise in marketing and operations, he helps businesses expand through building scalable systems, uniting teams, and generating demand. Ben has spoken on international stages, led full-day workshops, and worked alongside industry leaders around the world to educate entrepreneurs on cutting-edge strategies for business growth. Ben believes that by helping others win, he can elevate more people to live life on their own terms. When Ben is not working, he enjoys long walks in nature, reading, and a good board game.

Contact Information
Facebook: https://www.facebook.com/BenDahlMarketing/

CHAPTER 6

THE ONE SECRET TO LONG LASTING FITNESS

By Bernard Yeo

Stand at a busy intersection and observe the people passing by. What do you notice? A significant number of people are overweight. In today's world, the internet and social media provide information about health and fitness in an instant, yet the number of overweight people is increasing.

If you struggle with being overweight or lack motivation for exercise, I might be able to help.

Being in a comfort zone makes us feel safe and certain in routine activities and habits, so we may feel that stepping out of this zone could bring unexpected risk and stress. For instance, if you've never been the exercising kind, you'll need the motivation to get started and stay on course, right?

Most of us know someone who had the sudden motivation to lose weight, whether it was to impress a girl or get in shape for the summer or a planned holiday. But once the girl moves on to be with someone else or when the event is over, the motivation disappears, and the weight gain returns.

You see, these motivations are external influences. It's not necessarily a bad thing. It is effective for jumpstarting a fitness program. I used to think all it required was enough motivation to act and succeed in getting healthy and fit. I soon learned it also takes determination and willpower to stay motivated, which is why so many people give up after a while.

So, how do some people make keeping the weight off look so easy? How do they stay so motivated for so long? What is the differentiating factor?

There is another element more powerful than motivation. And it is not another kind of motivation. I often wondered why I could stay motivated long enough to be successful at certain things while some of my friends couldn't. Then, I came across research by Dr. Maxwell Maltz, a cosmetic surgeon. He noticed that after correcting imperfections on his patients' appearances, some still felt they were ugly, and this led to his discovery of self-image. In his book, Psycho-Cybernetics, he said, "A human being always acts and feels and performs in accordance with what he imagines to be true about himself and his environment." So, it wasn't what his patients saw when they looked in the mirror. It didn't matter because they considered themselves ugly.

Self-image determines the actions and decisions a person makes every day. For example, someone might have a goal to lose weight, but he continuously sabotages his efforts by overeating sugary foods. Why is that? His self-image is that of an overweight person and so his subconscious mind will consistently act in accordance with his self-image.

Effective weight management starts with motivation, but it won't sustain you. You need to program your subconscious mind and develop a good self-image to be successful in the long term.

Looking back on my life as an example, I started smoking in my early thirties. It probably started with friends over at my house. They wanted to smoke, and I wanted to be social. Soon I was taking smoke breaks with my work colleagues as well. But, after a few months, I didn't feel good about smoking. I felt that it wasn't me. Unbeknown to me at the time, it was probably Dr. Maltz's influence from the power of self-image. Perhaps my self-image of being a non-smoker came from my father. He smoked ever since I knew him and then he passed away from cancer when he was only fifty-one years old. Prior to discovering he had cancer, his engineering company still had hundreds of thousands of dollars owing to the banks for machinery leasing, loans, and recurring costs in operating a factory. My mother, my brothers, and I had to pick up from where he left off as it was our only source of income. It was terrifying because we didn't know anything about operating a business. Losing him was the hardest episode of my life and it was a shame that he couldn't teach us any business skills in time.

Not to brag, but I found it wasn't difficult for me to quit smoking as I had no withdrawal symptoms or nicotine cravings. So, despite the many people around me who smoke, I couldn't be influenced for long. While people normally fail to quit smoking, you can say that I *failed* in smoking!

Another significant event in my life was when I wanted to lose weight in December 2010. I was already in my late thirties and I had not exercised for years. While I was deciding on what to do, I found out that my cousin, Thomas, had just finished a full marathon. Ever since I was a teenager, I admired people who ran a full 42.196 km. I couldn't even run 5 km without stopping a few times. So, my cousin's success inspired me to get into running.

My idea of succeeding was to totally immerse myself in running. First, I set a goal to run a marathon in six months. Then, I bought a book on the beginner's guide to running a marathon. I started to train according to the plan in the book with almost no deviation. I watched proper running techniques on YouTube, I signed up for shorter races leading up to the big event, and I became aware of my food choices. I imagined myself running like Ancient Greek messengers delivering letters in times of war.

On marathon day, it was a tough and hot race. I had cramps in muscles I never knew existed. But I completed the race in five hours and one minute.

Over the next few years, I would run more marathons and even a 100 km ultramarathon. In 2018, I decided to shift from running to lifting weights.

In my marathon example, I was initially motivated to run a marathon in six months. I was motivated by my cousin's success. But, I also totally immersed myself in the sport, so much so that being a long-distance runner formed part of my self-image, which made me get out there to train, regardless if it was too hot or raining.

Now, we circle back to my original promise to you - to help you be successful in your fitness journey. I recommend you get started by motivating yourself and developing the self-image for long-lasting change. How long should you keep yourself motivated – until you achieved your goal and transformed your self-image.

Start your fitness journey with a reason. For me, it was my wife and children. I never want them to experience what I went through with my father's death. Write a letter to yourself (don't type on a computer) and tell yourself why you want to be fit. The letter can be any length, but the longer it is, the better. I want you to make a *promise* to yourself that you will achieve your goal. I say promise because breaking promises is more emotional than just missing a goal. I have taken out letters to myself written months ago and whenever I read them, I get emotionally charged.

Another powerful technique is to announce your goals on social media like Facebook. Letting all your friends know and holding yourself accountable can be incredibly powerful. You will get a lot of motivation and encouragement from your friends. Similarly, with your letter, on Facebook, you are promising your friends.

Build your self-image. Dr. Maltz said to change our self-image, we must use our imagination. By imagining a new healthier version of ourselves, our actions will be redirected to make it come true.

It can be hard to imagine because we are preoccupied with distracting thoughts, so meditation can be effective through implementing more of our senses to achieve our goals. So, here's what I want you to do: Take your self-image vision from the back of your mind and put it in front of your eyes by writing it out on paper. Again, there's something real about the act of writing things down. You employ the sense of sight, touch and hearing your written words in your mind. Create a vision of what you want to be. Put it on your desk or refrigerator and look at it everyday.

When You Don't Feel Motivated

Here's the other side to achieving your goals. There will be days where you don't feel motivated to do anything. You won't feel like going to the gym or for a run. I've faced laziness and procrastination too. There are some people who suggest dressing in a gym or running attire is motivating to get going. If you find an effective technique, then use it.

Here's what worked for me. When I was a runner, I told myself I'll only run for five minutes. If after 5 minutes I don't feel like running anymore, I'll stop. But, after five minutes if I feel like I want to continue, then I will run for another five minutes. Then I evaluate again if I want to continue or stop. I have run extra distances because of this mind hack. And, if your exercise is at the gym, tell yourself to go but only commit to do the warmup sets of your favorite muscle group, then decide if you want to continue the workout or not. You haven't failed if you stop after five minutes. Think of it as a sloppy day and let it go.

Another way to beat motivation (or lack thereof) is to integrate fitness into your life. Consider using resistance bands while watching television, set exercise time using calendar appointments, or have friends pick you up at scheduled times to go to the gym. Stick to these ideas and I promise you, you'll see the change in your self-image.

I want to close this chapter with the story about the scorpion and the frog. You may have heard of it, but I have a different view. The insert below is from Wikipedia:

A scorpion asks a frog to carry it across a river. The frog hesitates, afraid of being stung by the scorpion, but the scorpion argues that if it did that, they would both drown. The frog considers this argument sensible and agrees to transport the scorpion. The scorpion climbs onto the frog's back and the frog begins to swim, but midway across the river, the scorpion stings the frog, dooming them both. The dying frog asks the scorpion why it stung, to which the scorpion replies, "I couldn't help it. It's in my nature."

Just like it is the scorpion's nature to sting, you must establish a healthy self-image with fitness in your nature. If you do, you will never have to struggle with weight issues again.

BIOGRAPHY

Bernard Yeo is known for his love for personal development and dedication to personal fitness. Having run marathons for several years and switching to weight training, he has redirected his passion for coaching others on how to fight the battle of the bulge and gain fitness. With years of virtual traveling overseas and YouTube on his belt, he now makes actual adventurous and challenging trips with his ex-classmates every year. Bernard lives with his wife and two sons in Kuala Lumpur, Malaysia.

Contact Information
Facebook: https://www.facebook.com/BernardYeo
Instagram: https://www.instagram.com/justdoit.mindset/

CHAPTER 7

NEVER ENDING JOURNEY
By Carolyn V. Anderson

Success for me continues to be a never-ending journey of challenges followed by favorable outcomes. Pulling back the covers and getting out of my warm bed turned out to be the biggest challenge on this cold winter morning. I once worked for a successful attorney who told me that success started with waking up and getting out of bed every day with an attitude of gratitude. Years later, when I was going through the divorce process, remembering his words and enforcing that positive attitude saved me from falling into depression.

According to the dictionary sitting beside my computer, the word "success" has more than one meaning. Apart from gaining wealth and fame, it also describes success as achieving a favorable or desired result. The point is if you are happy with your outcome, you are successful.

For as long as I can remember, I liked sharing anything beautiful or unusual that I saw in the world around me. My father, an artist, taught me to sketch and reproduce my interpretation of the surroundings on paper.

I was immersed in sketching, doing it as often as I could find the time, thereby improving my skills. But the thought of making money from my art never occurred to me. In high school, an advertising agency collaborated with our school's art department and sponsored a poster contest with cash prizes.

The agency was in search of ideas for billboard advertisements for its clients. The design had to be simple but large enough for drivers and passengers in vehicles on the highway to see and read the message in a flash.

Our art teachers provided the poster boards and poster paint. I visualized my design and then sat down with paint colors to sketch and paint. When our posters were completed, the teachers submitted them to the agency's team for a decision.

On the day of the contest results, we all gathered in the school's auditorium. The first poster displayed on the stage was the poster for third prize. I recognized my poster and heard my name being called. I walked up to the stage with a bounce in my steps and a smile on my face to receive my envelope of cash.

As the second and the first prize winners received their envelopes, a sense of curiosity directed my attention to the other winning posters. I realized that even though I had submitted a well-thought-out and artistically-done poster, the subject of the other two winners was better aligned with the vision of the agency's clients.

First lesson learned—*always give the client what they are looking for.*

DEVELOPING SKILLS

I pursued my passion in college and majored in art and design. I also picked up new skills in architecture, blueprint layouts, and exterior and interior designs. In my oil painting class, our Instructor challenged us to use color to show line and movement on a new canvas. I had been up the previous night studying for an exam in my next class, so I chose to paint a subject similar to what I had studied. I created something different from my usual work. By doing so, I gave him exactly what he was looking for, and not surprisingly, that painting ended up in the university's gallery.

In one of my design classes, the Instructor gave us the taste of creating "light and dark" designs on photo paper. I was enchanted with the developing process in the darkroom. Seeing the images come to life in the developing trays reignited my passion for recreating beauty around me. It was a much quicker way to share images that would make others smile.

After college, as we stepped into the real world, only office desk jobs were available to women at that time. I was good at answering phones and could type, so a job downtown was not hard to find. I started looking for photography classes close to work as I did not want to put my new passion on hold.

I learned to be aware of the lighting and contrasting color shades while taking black and white photos. Acquiring new tricks of developing photographs, I experimented with my grandfather's 35-millimeter camera with a retractable "billow." I took photos during my lunch hours and sometimes after work.

Grateful that I had a salary and paychecks to rely on

Such was my passion that I started saving money to purchase equipment for developing photos. It wasn't long afterward that my closest friend and I made plans to fly to the New York World's Fair later that year. I was still living at home when my parents decided to move closer to my grandmother. As I wanted to move with them, I purchased an automobile to commute to work. In an effort to save money, my girlfriend and I drove to New York from Michigan in my new Chevy. With a newly-purchased Kodak Instamatic camera, which automatically advanced each frame after the shutter closed, our journey began.

I had enjoyed traveling to several states and Canada before college, so this was a much-anticipated trip. We drove straight to our hotel in New York. Parking my car in the hotel's garage, we relied on the subway system to get us to the fair and back. Everywhere we went, my camera accompanied me. I gained an appreciation for the quick shutter movement when we took the "ski lift" above the fairgrounds from one end to the other. I had a lot of photos developed by someone else and put together a photo album to show to family and friends who otherwise would never see the World's Fair.

Another detour before building a darkroom

I had finally saved enough money to purchase the necessary equipment for my darkroom. But by the time the equipment arrived, I had fallen in love and was planning my wedding. So, it went to storage at my parent's house.

I hadn't been to San Francisco since I was seventeen. My husband's friend lived there, so we flew to San Francisco for our honeymoon. I carried one of my cameras along on our tours in and outside the city and visited sights that I missed on my first time there. On the weekend, we flew down to Los Angeles to visit my sister-in-law's husband. She had come to our wedding, but her husband could not get off work to travel. It was the perfect opportunity to capture the city in its full spirit. The next week, we drove to San Diego—I had never been there and was eager to take photos by the water. A few days later, we were homeward bound flying out of the Los Angeles airport. Yet again, I sent the film out to be developed, but at the back of my mind, I was dreaming of my darkroom.

We rented an inexpensive apartment, which allowed us to save for a deposit on a home we eventually purchased. Our firstborn was five months old when we moved into our new home. There, I set up a darkroom and went to work in it, but it proved difficult. When our second child was born, it became almost impossible to make use of the darkroom. Remembering something I heard long ago, "Out of something negative comes something positive." I realized I still had my camera and knew where I could have my film developed. I was grateful to watch my children grow up and capture those precious moments on film.

When my children were in school, I opened up my darkroom. The photographs I developed were mostly shared as gifts; that would put a smile on the recipient's faces. The success I achieved could not be evaluated monetarily.

Several years later, in the process of divorce, I could no longer afford the cost involved in photography. I had a son and daughter in their mid-teens and a job that barely paid the bills. The photos and processing were not paying for themselves, so I had to shut it all down.

The thought of starting afresh and reopening my darkroom began to take shape when my children ventured out on their own. But without a plan, my enthusiasm for the process wavered. Out of the blue, one of my girlfriends asked me to share a booth with her at a local street fair. I created greeting cards to match her crafted door hangers—this plan ended up being mutually beneficial. So, we made plans to do it again the following year, and my daughter encouraged me to sell my photos this time.

As there were several craft booths selling photos, I decided to place my black and white photos on greeting cards. A friend of mine owned a printing shop with her husband, so I asked her for help. She showed me how to do the layouts, which they ran through the press on heavy stock paper. Then they scored the seam to be folded, and it was ready for the next step. I purchased envelopes for the cards and applied for a copyright. Grateful for her help, I suggested she keep a few cards to display as samples for their customers.

At the last moment, my booth partner backed out. With the help of friends and the women in the two booths around me, I set up my display. Pricing the "photocards" as single cards, I also sold them in packages of ten. I helped the two women next to me by watching their booths whenever they needed a break. They did the same for me. By the end of the two-day fair, the three of us had become friends. For the next four years, we requested the same spaces next to each other.

Battling the rainy clouds, the sun shone brightly on the day of the craft show. I was pleased with the successful sales of the cards. Michigan Lighthouse photoprints drew the most attention—coincidentally, they were my favorite. Now a new journey was materializing; to find other lighthouse enthusiasts and join them in visiting and photographing lighthouses.

I realized that success achieved through working with others is undoubtedly more fun. Finally, I had found an opportunity that made a positive difference in the lives of others. I discovered that as a team, we reached more people, identified their needs, and helped them find solutions. It was a whole new level of success beyond our expectations. When you only focus on personal success, you limit yourself. When each member of the team develops another person, and that person develops another who, in turn, develops the next, and the next, you will have a chain of people, a team, working toward a common goal.

I have often read that "You are the average of the five people you spend the most time with." Often visiting The Henry Ford Museum, old photos of three successful men come to mind—Thomas Edison, Henry Ford, and Harvey Firestone, who all shared a strong friendship. Surround yourself with the right people and build a network of successful people as your support group. The results may surprise you.

BIOGRAPHY

At a time in Carolyn's life, when most people retire, she is infused with a strong impulse to see what is around the next corner. She hopes that her story inspires you to pursue your dreams with passion despite the challenges life throws at you. She believes that "You only fail if you quit." She acquired her writing skills with the help of her high school journalism teacher. Carolyn recently joined other best-selling authors in *The Art and Science of Success, Volume 5,* with her chapter, "Finding Success Every Day." She learned teamwork and leadership skills through local and national organizations: parent-teachers association, 6-Area Coalition for Community Mental Health Board of Directors, and Parents Without Partners (as a program and education leader).

Contact Information
Facebook: https://www.facebook.com/CarolynsVerve

CHAPTER 8

FACE YOUR GRIZZLY BEAR
By Chris McIntosh

You are squatting on the bank of a snowy stream, filling your water bottle. As you watch the water flow into your bottle, a thought engulfs your mind—you will be forever lost in this unforsaken wilderness and die. You will never be rescued. Then, images of your family back home come flooding in. Your children need you to teach them about life. They are counting on you. And they have no idea where you are at this moment. Something kicks in, and you begin to strategize your way out of your situation, blaming yourself for getting into it in the first place. You hope to get rescued, for everything you have tried so far that didn't work. As your belly reminds you that it has been too long since your last meal, you rack your brain to find food and survive another night without freezing to death. You look down and see your tired reflection in the stream and think, "If I follow this, maybe it will show me a way out."

As your water bottle is almost full, you hear a noise in the bushes, upstream from where you are squatting. You turn, look up, and your eyes lock with a grizzly bear, who is now casually strolling toward you. The bear you thought you had escaped from days earlier, has found you again. Now, he is approaching you directly, and he is not coming to be your friend. Fear suddenly consumes you and dropping your bottle, *you run!*

We are all created with a purpose, destined to do great things in our lives. But why is it that most people lead what Henry David Thoreau called Lives of Quiet Desperation? For most people, the biggest regret at the end of their days is not the things they did and failed at, but those they were too afraid of doing. Not speaking up for themselves, not doing *that* thing they always wanted to do, not approaching the person who made their hearts flutter, or not making amends with a family member constitute some of those regrets. People regret being risk-averse and not starting that business they had always dreamed of; not taking enough vacations; not living the lives they were destined and created to live; of not having the courage to live lives they wanted instead of the lives others expected them to live; and of working hard all of their lives only to build someone else's business and someone else's dream.

Yet again, you have gotten away from the grizzly bear. But this time, you realize that it is only temporary—he will find you again. As long as you are in his world, he will sniff you out. You are no longer thinking about being rescued, about getting to a place of freedom. All you can think about is how this grizzly bear is going to take you out, and when it will happen.

As you're contemplating your demise, your thoughts revert to your future, to your family, to the things you still need to accomplish in this world, and at that moment, you make a decision—I must face this grizzly bear; I must kill this grizzly bear. How can you ever do that when all you have is a small pocket knife? Surely, a pocket knife could never penetrate deep enough into the bear's hide to do any lethal damage, but then you remember reading about how others had slain grizzly bears with nothing more than a spear.

We all have so much potential to achieve great things in our lives, but we've built these invisible and internal barriers. We are a powerhouse of so much information, skills learned through the books we have read, and from the life we have been living. Courses bought, money invested in the boot camps—yet we still do not have the success we deserve.

Why is it that success evades you? Is it because you didn't buy the right course, or you don't have the right information? Perhaps you didn't read the right book, or you just weren't born into the right family. The answer to all of these questions is "no." You have not yet arrived because there are grizzly bears in your life. Each time you begin to tread on the path of success, the grizzly bear is right there, waiting for you and ready to devour you. Ultimately, fear grips you, and *you run*!

One of the most popular acronyms for fear is "False Evidence Appearing Real." I have heard another one—"Forget Everything and Run." Fear is a powerful emotion and a primal instinct, which compels one to *run* in the face of danger. It consumes you, and if you let it control you, it ends up controlling your life. Eventually, you begin to justify the fear and consider it a new norm. "Living by default"—our whole lives, we are programmed to live a certain way. The problem with a default way of life is that it is just too easy. It is too easy to show up every day at the job you hate. But we justify it by saying, "At least, it's a paycheck." We are programmed to go to school and get good grades so that we can go to college and receive a degree (along with a mountain of student debt). And why do we need a degree? To get a job, of course, and they pay you just enough so you won't quit, and of course, you work just hard enough so you won't get fired. Besides, "they have great benefits," you say. I call this "the golden handcuffs."

Now, there's nothing wrong with getting a degree and having a job. But not having a plan to get out of that job when you want, is wrong. The greatest country in the world, the United States of America, is the most prosperous country in the world. People risk everything, including their own lives, to get there. Then why is it that even at the age of sixty-five, nine out of ten Americans will be living at or below the poverty line? Because living by default is easy. Even right now, as I write this, the economy is booming, yet many Americans will come up short when it comes to retirement income. The idea of being a greeter at a Walmart, when you should be enjoying your golden years, is not exactly enticing. According to the Pension Rights Center, half of all Americans at the age of sixty-five or older have incomes of less than $24,224 a year—far less than the amount most need to meet their day-to-day living and health care expenses. Yet, I am grateful for companies like Walmart that are willing to employ the elderly in our society who fall in the trap of a "default life."

"Face Everything And Rise"—my favorite acronym for fear. We all have grizzly bears or fear in our lives, but how we deal with that fear determines our future. The only way to design your life instead of living by default is to first slay the grizzly bear in your life. I am no different: I bought the books, the courses, invested in boot camps, but I never had any success. That is until I realized that my success had nothing to do with the information I had purchased. Merely hoarding information or knowledge wasn't enough; I had to take action. Most people's inability to take action reflects their deepest fears, current behaviors, current beliefs, and operating norms. Their assumptions and rigid patterns prevent them from living by design, and they resist any change in their lives. I once heard that the only person who likes change is a baby with a wet diaper. I was at a place in my life when I realized that I had grizzly bears preventing me from making change, and harvesting my true potential.

I had humble beginnings, so I needed to overcome the invisible barriers to achieve success. As a young child, I often saw my mother rip pieces of paper called food stamps out of a book, as we didn't have enough money to buy groceries. The embarrassment of standing in line in that grocery store is still fresh in my mind. My father refused to go with my mother as he could not bear the humiliation. I don't have a college degree—my parents could hardly afford food, much less pay for college. I'm the oldest of six kids. My dad was in the timber industry—he was a logger, or as some people call it, a lumberjack. I love my mom and dad, and they did the very best they could with what they had. Unbeknownst to them, they lived a life of default. Our broken educational system fixates on how to score a job. To my parents, they did everything right—my dad was and still is a very hard worker. After the Vietnam War, my dad did the honorable thing that every father does. He went to get a job to provide for his growing family. He didn't have a college degree either, so he did exactly what his dad had done, and that was to go to work in the timber industry. And he taught me how to do the same—to be a hard worker—and that's what I did. I worked hard, and I thought that the harder I worked, the more ahead I would get. So, I followed after my father. My dad did give me some advice, though. He said, "Son, don't go into the Army and learn how to be a mercenary. Instead, go into the Air Force so you will at least learn a skill. So, you won't have to work in the woods your whole life."

Following his advice, I went into the Air Force. Please, don't get me wrong—I am grateful for all branches of the military. It is because of our military that our country is as great as it is, and we have the freedom to choose to live by design. I would recommend my boys to serve in the military, as you can learn so much about yourself and discipline from the armed forces, but I also teach my children to live by design.

We all have defining moments in our lives. Reading this book could be your defining moment. I encountered mine during a conversation with a letter carrier. After getting out of the Air Force, I held various jobs, including working in the woods as a lumberjack. While working as a letter carrier at the post office, I met another letter carrier who had been working at that post office longer than I had been alive. At that point in time, I thought I had finally arrived. I had a government job—there was no way I could get fired, and the pay was really good. When I found out that this letter carrier had been delivering mail longer than I had been alive, I was blown away. I was in my early thirties, and the conversation had come to the place where I asked him how much money he made annually.

That question and his response was my defining moment. When he told me how much he earned, I realized that I could no longer work for the post office. At that time, I was earning more money than him on an annual basis. How could I be making more money, having only been at the post office for a short four years than someone who had been there for thirty-plus years? I blindly followed what my dad had taught me, and I worked hard.

> *"If you don't find a way to make money while you sleep, you will work until you die."*
> –Warren Buffett

I was trading time for money. I worked so hard that I requested to be on what was called the ODL, the Overtime Desired List. This concept may sound foreign to some: I worked ten to twelve hours a day, six days a week, and thus, I made more money on an annual basis than a thirty-year US postal worker who only worked a forty-hour week. But as is customary, my standard of living had adjusted to the money I was making, and I was not getting ahead financially. It hit me that I was not willing to work ten to twelve hours a day, six days a week for the rest of my life. I had a young daughter at home, Makayla, I adored her, and I wanted more children. I remember that I didn't have a lot of time with my dad because he worked so much, but I treasured what little time I had with him, which was usually only on the weekends. I cherish the time that we spent hunting, camping, and fishing together. I fondly recall that on weekends he would take me with him to the gym, and we would play a game of pickup basketball with all of his friends. I wanted to spend more time with my family and wanted that for my children, so I quit the post office to pursue a business that I hoped would get me to the land of financial freedom. Soon after, the grizzly bear came. Things got bad. I got divorced. Life turned upside down.

In the movie, *The Edge*, which I highly recommend you watch, Anthony Hopkins is being chased by a grizzly bear. He realizes that his only option is to kill the grizzly bear. It is the statement that he makes in the movie that changed my life, as I hope for this book to change yours.

He says, "What one man can do, so can another." I had long realized that I was destined for success in my life; so, all I needed to do was to learn what others have done to find success and then replicate it. I wanted to leave a legacy for generations to come, so investing in real estate made sense. For me, real estate investing was the vehicle through which I would create a passive income and free up my time. Over time, I learned what others had done to gain success, not just in real estate—because if it was easy, everyone would be doing it—but in life.

We all have grizzly bears that chase us down and prevent us from taking action, from getting to that place of ultimate success, and it is very scary. Your choice, whether you *Forget Everything And Run*, or *Face Everything And Rise*, makes all the difference. Remember, you do not have to carry the burden all by yourself: "What one man can do, so can another." Face your fears, slay your grizzly bears and change your life. Stop living by default and instead choose a life of design. If you won't do it for yourself, do it for your family, for your legacy. Is it scary? Yes! But you can do it. I believe in you.

BIOGRAPHY

Chris McIntosh has been an entrepreneur for over fifteen years. He has coached, mentored, and inspired hundreds of entrepreneurs and real estate investors over the last decade. He founded and is currently the president of the longest-running real estate investing club in his hometown of Spokane, Washington. He manages a multi-million-dollar real estate portfolio and helps others get into the game of business ownership and real estate investment. His mission is to inspire others and empower them with universal truths to lead a life of design. He is the proud father of Makayla, Austin, Noah, Caleb, and Jessica.

Contact Information
Facebook: https://www.facebook.com/mcintoc
Instagram: https://www.instagram.com/tip.real.estate/
Website: http://www.nextgenrei.com

CHAPTER 9

OVERCOMER

By Cindy Cavazos

I remember from a young age having to take care of my brother, who was a year younger than me, as well as a three-year-old child. At the time, it was just my brother, my mother and I; I didn't have a father in my life and knew nothing about him. The only other family I had was an aunt. My mom was single who often worked nights, so I was left alone to care for my brother and anyone else who was with us, and usually, instructions were left for me to warm or cook dinner. We must have been on food stamps because I remember the book of play money. We also moved around often, so I don't have any childhood friends.

One night, while my mom was at work, someone broke into our apartment. I was lifted from my bed and taken to another room. I woke up with a pillow over my face hearing whispers, while a cold metal object was pressed to my forehead. It was a gun. I couldn't see the man, but he was molesting me. When my mom came home, she found her gun had been taken with other valuables. I said I was okay after the incident, and it was never mentioned again. I was eight or nine years old at the time.

Later, when I had children, I did not leave them home alone until they were eighteen.

When I was nine, my mom met a Greek man, and they had a daughter, my sister, who I took care of since she was two weeks old. I remember staying up all night when she had colic. My mom was at work, and I had school the next morning. It was just our routine. My brother and I looked after our sister for many years while my mom worked. I recall sitting on the door frame of our 2-bedroom apartment in Houston with my siblings as the sun set because the light between our apartment door and neighbor's door was the only light we had. It wasn't a power outage. The light bill hadn't been paid. I think a neighbor called my mom to let her know our lights were out.

Throughout my life, my mom worked 2 to 3 jobs to make ends meet. She missed out on holidays, our birthdays, including her own, as well as Mother's Day. She also missed our school functions. During a school production of Hansel and Gretel where I played Gretel, I remember standing on the stage looking for my mom and nearly forgot my lines while looking for her in the ocean of faces. She wasn't there.

My brother and I were due to go on a field trip, and although I don't remember if it was through the church or school, I do remember asking my mom repeatedly to wake up and drive us there. We were late, and they were going to leave without us, but she couldn't wake up. When I came back into her room, she was trying to place her legs through a paper bag, thinking they were pants because she was still asleep. We eventually managed to wake her, and we made it to our field trip, but I will never get that image of my sleep-deprived mom out of my mind.

It was sometime in junior high when my mother met my second stepfather, who was all smiles at first. He worked and helped my mom with bills, and they too had a daughter, my youngest sister. He showed my siblings love, but he was addicted to drugs and alcohol. To this day, I don't know the substance he was taking. I just remember the syringes and glass pipes.

The abuse was subtle at first, which then started escalating from corporal punishment to verbal abuse to outright hitting me. It was during high school, either as a freshman or sophomore, when my relationship with God blossomed thanks to two young school friends. They invited me to a church where I started building a relationship with God who I adopted as my father since I didn't have one. On one occasion, my friends came to pick me up for church and my stepfather said, "The bitches are here for you." My stomach turned. How could he call the people who shed love and light into my life such foul names?

He had been drinking straight for days and had lost his job. One night he came into my room and hit me for no reason while sitting next to my sister. I called my mom and he denied hitting me. A screaming match followed and he threatened to kill us. He then left the house, and I told my mother I wasn't waiting for him to return to kill us, so I was leaving. She packed our clothes and we all left for a friend's house, where we stayed until my mom got a new apartment. After months had passed, my mom told me that my stepfather was coming back to be part of our family. I told her that if she chose him over us that I would leave when I turned eighteen; I was only sixteen at the time. She picked him.

We moved so much during my childhood that I couldn't tell you how many schools we attended. I asked my mom to let me go to the same high school for all four years, but she broke that promise when she decided to move our family to Florida for a business venture that summer. I left the few friends I had where I managed to be part of the drill team and played varsity soccer as a sophomore. Now, all of that was over and I had to start new. I tried to stay behind with my church family but my mom wouldn't allow it. I cried the whole drive to Florida.

Even though my stepfather had stopped drinking, he was still verbally abusive. I wasn't even allowed to have an open-air vent in my room; he would come into my room and close the vent. So one time, I removed the whole vent from the wall, and he ransacked my room looking for it. I started working as a cashier at the local grocery store and I needed to wash my work apron frequently because it got soiled easily with milk, raw meat, fruits, and vegetables. Yet, I was only allowed to use our washer and dryer once a week. Another time, I didn't wash the dishes properly at home and my mom came to my work to look for me. I hid from her and didn't go home that night. I stayed with aunt instead. Eventually I returned home.

I stayed busy, filling my time with school, work, and gym. There was a law school close by that I contemplated attending after graduating from high school. I asked my mom if I could, and she told me to ask my stepfather, the same man who bullied me daily. I knew then that I had no future in that household. I saved money to return to Houston and start my life there. I turned eighteen in the June before my senior year and told my mom I already purchased a bus ticket back to Houston and that she could not stop me from leaving. My stepfather and my mom sat me down, and he confessed and apologized in front of my mother, saying he had purposely made my life miserable. I believe this was my mom's attempt to get me to stay, but I knew where her loyalties were, and they weren't with me. I left my siblings and my mother, but I was ready to live a life without abuse.

That summer, I lived with a church family back in Houston, and circumstances made it difficult for me to enroll in school and finish my senior year. It was not easy to have my school records transferred as I needed the paperwork I didn't have. One day my church family said they were moving and no longer had room for me. I found my stuff in a bag at the church when I returned from a youth church retreat and found that my money had coincidentally run out too. I would contribute financially and do things like take the family out to eat and help pay for car repairs if they didn't have money, so I felt they used me until I was left penniless. I didn't have a car or a bike, so I had to depend on other people to give me rides. I asked a church friend if I could live with her; she was staying with her boyfriend's family. This arrangement didn't last long, so we both stayed with her mom who was living with another family. I didn't have any relatives in Houston, just church family. I moved from home to home and finally decided to go live with my grandmother in Monterrey, Mexico to try finish my senior year there. I applied to the local school, but they needed my school paperwork. I begged my mom to send it, and although she said she did, the papers never made it to the school. I worked as a teacher and a telephone operator to offset expenses and I took a bus everywhere I needed to go.

Meanwhile, my mother moved back to Houston. I wanted to finish my school so badly that I moved back to my mom in Houston, despite the fact that facing my stepfather would make me ill. With my stepfather's presence, it continued to be an awful living situation. I got a job in the same mall as my mom so we could share rides to and from work, but my stepfather said that I had to find my own ride. He would leave me there on the sidewalk, stranded at 10 pm, as he drove away with my mom. There was no public transport system anywhere near there, and I was forced to ask strangers for rides or call taxis. Day-to-day life was challenging.

My stepfather wanted me to start paying rent. One evening, there was an argument, I don't even remember what it was about, but I just packed a small bag and started walking. I had a high school friend pick me up and I lived with him. We became romantically involved but it ended after a phone call between him and another girl where he denied my existence, while I was living there. We fought and he threw me out. I didn't have anywhere to go; I was only eighteen.

I looked in the Yellow Pages and found the Covenant House in Houston where he dropped me off.

The shelter wanted me to take a GED course, but I refused and went to the local high school where the school counselor wrote a letter to the shelter stating they would allow me to finish my school year. The shelter let me continue living there while I went to school, but it wasn't easy. I had to share a room with three strangers. I had hundreds of dollars, clothing, and sentimental jewelry stolen from me. I walked into my room once to find a guy shooting up drugs. I worked, went to school, and lived in the shelter, which had very strict rules. Some of the residents could not adhere to the rules and were either kicked out or left of their own accord. Some of them, male and female, were consumed by the streets of Houston and turned to prostitution. I cried every night that I was there.

Another church family reached out and explained they were willing to cosign an apartment for me. After living in the shelter for three months, I was able to leave and lease my first apartment with the cosigner's help. I started working as a waitress nearby and managed to pay all my bills on my own. I was still going to school to finish out the year and graduate, but to do so, I had to take two buses from southwest Houston to the inner-city high school, which took over an hour. Then, a family friend gave me a '78 Chevrolet Monte Carlo to drive to school and work. It was an old car that was expensive to maintain, but it was still better than waiting for a bus in the rain. To finish my high school education, I had to take a night school class at MacArthur High School on the northeast side of town where there were regular drive-bys and high gang activity. I fitted right in with my '78 Monte Carlo and felt like I was in East Los Angeles.

I graduated from high school with an honors diploma. I didn't attend graduation because I didn't know anyone in the school and knew nobody who would come support me. After the first six months of living alone, I was able to transfer the lease for my apartment into my name solely. The boyfriend I lived with before the shelter, moved in with me and that was the beginning of a 20-year abusive and loveless relationship. I went from waitressing to working as a leasing agent in the office of the apartment complex where I lived. I took a year off from school to figure out what I wanted to study. The Monte Carlo engine died on me around the same time my stepfather did the same.

My mom called to tell me that my stepfather came home and was throwing up blood on the driveway. He was hospitalized and given a few days to live. My mom requested I go to see him, but I didn't want to; however, she urged me to go and I eventually relented. When I entered his room, he immediately sat up in bed, moving his arms and legs up and down uncoordinated. Obviously, he wanted to say something to me or even hit me one last time, but he couldn't utter one intelligible word. He died a few days later, and my siblings and mother were left alone once again.

I wanted to become an architect and decided to study electrical engineering at ITT. I worked nine to five and then attend school six to ten, Monday through to Friday. I was at the top of my class in college but in 1998, I became pregnant with twin daughters--one of the biggest blessings from God. I didn't think it was logical for me to pursue electrical engineering anymore because I was not going to put two babies in daycare. My daughters were born healthy. Their father and I decided I would continue to work because I made more money and had more benefits, and he would stay home with the twins. But, after about a month, I was ready to quit and be home with my daughters. We knew if we wanted to provide a future for our kids, we would either both have to work or move to Mexico where life was cheaper. Well, we saved, packed up and went to Mexico. I stayed home with the 2-year-old twins while their dad worked as a schoolteacher and gave private English lessons on the side. Yet, we still didn't have enough for food. I rented out our PlayStation in 30-minute intervals, then added snacks and drinks, but even then, the profit was just enough for us to eat. We sold our belongings at flea markets; we did what we could to put food on the table. God never let us go hungry. When we literally didn't have a dime to our name, a neighbor would come by and offer us a plate of food.

We had a very loyal customer who was ten years old, and I could always count on him to rent the PlayStation for 30 minutes every morning so that I could buy milk or eggs for the twins. It was interesting as this child was bullied because his mother would pick through everyone's trash to recycle paper, metal, plastic, and glass. He had nine siblings and his mother put the older children through college, and they were now lawyers and engineers. Even though he was bullied, he had more money in his pocket than any of the other kids, and they would often borrow from him as well as bully him. I favored this child. I treated him like my son and hated leaving him. My mom had put a bug in my ear about having another child soon, so I prayed that God would bless us with a son. I specifically asked for one boy, not two; otherwise I would have to sit on the floor of the car because we wouldn't have any more space for two extra car seats! Within three months, I was pregnant.

After the tragic event of 9/11, we returned home, and in the following month, our family was back in Houston. We were living with my husband's mother and brother, and that's when I began to realize where his loyalties lay. I tried to leave with my daughters, but his mother convinced me to stay. I shouldn't have. I should have walked away and never looked back. That's when I knew I had to make enough money to support myself and my children. I started the college enrollment process, but I didn't want to start until my son was a year old. My friends and family discouraged my return to school and discouraged me from becoming a nurse because of how hard it would be with three children.

I finished nursing school in May 2006 and graduated with honors. I worked hard as a nurse and encouraged the father of my children to go to school. He began his college education shortly afterward, but it took him six years to finish, during which time he did not work. After he graduated, he couldn't find a job. No one would hire him. I was so blinded, I made excuses, but ultimately, I didn't know any better. I didn't have a father other than God. I didn't have uncles or any good man in my life, just abusive stepfathers. I didn't see the red flags even though everyone who knew us did, and they tried to warn me. When he choked me for the first and last time, I moved out but I didn't file for divorce because I had done that before and lost a lot of money.

In 2015, I filed for child support, and when he found out how much he was going to pay, he contested custody in the Attorney General's office. To afford a lawyer, I started working more and more, but I was tired of working weekends and holidays. The hospital owned me. I spent more time at work than at home with my kids. I barely saw my kids. I remember at one point, I was on the edge of my bed crying on the phone to a friend about not wanting to go back to work. Then, one day in September, everything changed when a good friend called and showed me this amazing concept. I only had enough money in my bank account for rent that was due in ten days, but I knew that if I didn't do something different, then nothing would change. Knowing that God was placing something very important in my lap, I took a leap of faith, and this concept has been a blessing in our lives ever since.

In June 2016, my ex took my son and refused to return him. When I went to pick my son up, we both called the police. He told the police I assaulted him, but I never put my hands on him. On the contrary, he was the abuser. The police believed him and arrested me, but they couldn't charge me with assault because I never touched him, so they charged me with burglary of my own home. There I was, handcuffed in front of the house I built for my family and my twins in the Honda minivan I bought with my nursing salary. Yet, I had plenty of opportunities to have this man arrested, this man who lied to the officers and had me in handcuffs. Not only was my freedom in jeopardy but so was my nursing license. The same man I put through college for six years was taking away my freedom, my livelihood, and my children all on a lie.

I was treated like an animal in jail and wasn't allowed to use the bathroom for over five hours. I was later transferred to a maximum security prison where women were kept in glass cages. One of them asked me why I was there, and like in the movies, as I answered what my charge was, she finished my sentence. When I asked her how she knew, she said it was because she was in for the same allegation. I found out from her that she had been in for eighteen months, but if convicted, a person could serve two to twenty years. My knees buckled and my heart stopped. I could not believe what I heard. I did absolutely nothing wrong. I prayed harder than I ever prayed before and cried like I had never cried before. But God had a plan. That night, he gave me a vision that I would be reunited with my three children, and I saw us on a beach with blue waters and white sand, laughing and having fun.

I held onto this vision through nearly three years of court dates. I paid $900 to get out of jail, and thirty days after my arrest, my lawyer called and said the District Attorney rejected the case. Although I didn't have to go to court nor report to the board of nursing, the false arrest would still negatively affect me. My ex used this false arrest against me, which ultimately cost me more than my nursing license. The judge took almost all my parental rights away over my son. I was devastated and hurt beyond belief. I had felt no greater pain in my life. I paid five lawyers in total and also had to pay child support for my son even though I was taking care of our twin daughters. I worked almost every day in 2017 and only took off Thanksgiving and Christmas.

In June 2018, my mom called to ask me to move my holiday Tupperware out of her storage. I asked her jokingly if she was moving someone else in, and she said she was foreclosing on her 20-year home. She hadn't been working as she was taking care of my 99-year-old grandmother, whose caregiver of fifteen years had recently quit. I was crushed that my mother was losing her house. I couldn't believe I lost my home to my ex, and my mom lost her home to foreclosure. I swore I would never let anything like this happen again. In September 2018, I moved my mom and my grandmother in with us.

In early January, my friend Johnny called to invite me to a national training event that greatly impacted my life. I started building my business again. I ranked up and started earning residual income, as were my partners. I traveled more in those last six months than I did my whole life. I requested a meeting with my son and the judge accepted because my son wanted to spend more time with me, but my ex denied him that right. The judge finally interviewed my son and I've been granted additional visiting time ever since. I'm glad I never gave up when my friends and family told me I was wasting time and money on trying to get more time with my son.

My mom is one of the greatest women I know. One thing she taught me is that you can do anything you want to do. There is nothing that can stop you. God has given us the choice of right or wrong, and faith or doubt. Without God, without faith, I could never conquer the obstacles. I would have never survived my trauma. Ask, Believe, Receive, and Expect it. If I hadn't gone through the trauma I went through, wouldn't have a story to help others conquer their own obstacles! On many occasions, the following verse has reminded me of my path.

"But they that wait upon the LORD shall renew their strength; they shall mount up with wings as eagles; they shall run, and not be weary; they shall walk, and not faint," —Isaiah 40:31 King James Version (KJV).

BIOGRAPHY

Cindy Cavazos has been a Registered Nurse for more than fourteen years in Houston, Texas. She obtained a national Critical Cardiovascular certification and is the single mother of three children, her 20-year old twin daughters, and a 17-year-old son. She homeschooled them all their lives, and now the twins are in their second year of college. One twin has a 3.4 GPA and the other twin has a 4.0 GPA. Cindy makes over $100k annually. She sacrificed family time and was unable to help family and friends achieve greatness because of her nursing job. In the last four years, Cindy built a small Network Marketing Business that now has a presence in over three countries, and the business growth is allowing her to spend more time with family and friends. She wants to help a billion people and change the world with this concept.

Contact Information
Facebook: https://www.facebook.com/cindy.cavazos.35
Instagram: https://www.instagram.com/p/B1k2SBKBEAK

with my three children, and I saw us on a beach with blue waters and white sand, laughing and having fun.

I held onto this vision through nearly three years of court dates. I paid $900 to get out of jail, and thirty days after my arrest, my lawyer called and said the District Attorney rejected the case. Although I didn't have to go to court nor report to the board of nursing, the false arrest would still negatively affect me. My ex used this false arrest against me, which ultimately cost me more than my nursing license. The judge took almost all my parental rights away over my son. I was devastated and hurt beyond belief. I had felt no greater pain in my life. I paid five lawyers in total and also had to pay child support for my son even though I was taking care of our twin daughters. I worked almost every day in 2017 and only took off Thanksgiving and Christmas.

In June 2018, my mom called to ask me to move my holiday Tupperware out of her storage. I asked her jokingly if she was moving someone else in, and she said she was foreclosing on her 20-year home. She hadn't been working as she was taking care of my 99-year-old grandmother, whose caregiver of fifteen years had recently quit. I was crushed that my mother was losing her house. I couldn't believe I lost my home to my ex, and my mom lost her home to foreclosure. I swore I would never let anything like this happen again. In September 2018, I moved my mom and my grandmother in with us.

In early January, my friend Johnny called to invite me to a national training event that greatly impacted my life. I started building my business again. I ranked up and started earning residual income, as were my partners. I traveled more in those last six months than I did my whole life. I requested a meeting with my son and the judge accepted because my son wanted to spend more time with me, but my ex denied him that right. The judge finally interviewed my son and I've been granted additional visiting time ever since. I'm glad I never gave up when my friends and family told me I was wasting time and money on trying to get more time with my son.

My mom is one of the greatest women I know. One thing she taught me is that you can do anything you want to do. There is nothing that can stop you. God has given us the choice of right or wrong, and faith or doubt. Without God, without faith, I could never conquer the obstacles. I would have never survived my trauma. Ask, Believe, Receive, and Expect it. If I hadn't gone through the trauma I went through, wouldn't have a story to help others conquer their own obstacles! On many occasions, the following verse has reminded me of my path.

"But they that wait upon the LORD shall renew their strength; they shall mount up with wings as eagles; they shall run, and not be weary; they shall walk, and not faint," —Isaiah 40:31 King James Version (KJV).

BIOGRAPHY

Cindy Cavazos has been a Registered Nurse for more than fourteen years in Houston, Texas. She obtained a national Critical Cardiovascular certification and is the single mother of three children, her 20-year old twin daughters, and a 17-year-old son. She homeschooled them all their lives, and now the twins are in their second year of college. One twin has a 3.4 GPA and the other twin has a 4.0 GPA. Cindy makes over $100k annually. She sacrificed family time and was unable to help family and friends achieve greatness because of her nursing job. In the last four years, Cindy built a small Network Marketing Business that now has a presence in over three countries, and the business growth is allowing her to spend more time with family and friends. She wants to help a billion people and change the world with this concept.

Contact Information
Facebook: https://www.facebook.com/cindy.cavazos.35
Instagram: https://www.instagram.com/p/B1k2SBKBEAK

CHAPTER 10

SELF-IMAGE AND MY IDENTITY

By Devon Kurz

Self-awareness is knowing where I am in life, how I feel about it, what I am doing, and where I am headed. Who I think and believe I am is not as important. It is detrimental to my growth and the impact of presence to believe in who I am not. I must change the way I see myself and have an understanding of my true identity—who I am, the purpose for which I was created—as uncertainty breeds anxiety. This involves understanding who I am, who created me, and what I believe in. Do I believe what the world tells me I am, or do I believe it is my creator? I have struggled with this for the majority of my life…until now. I came to an understanding, a belief, after many years of fighting, pushing my body to its limits, working my life away, and thinking that I would get what I wanted as long as I worked hard and long. So, I took twenty-plus years, accomplishing many things except for the things most important in life: family, friends, marriage, and perhaps what is most precious, my beautiful children (Sierra and Conner), who are now young adults. I lost so many precious years and missed out on crucial developments in their lives as they grew up. I didn't see myself as the loving and caring father I wanted to be for my kids. I saw myself as the father who worked, providing money and not time. My identity was that of a workaholic and not the loving father I desired to be. After all that had transpired, I lacked any healthy relationships, I was in poor physical condition and health, and I finally came to realize that my life had to be different. I needed to change, though I didn't have the answer as to how to go about it. I didn't have a good mentor in my life—if I did, I wasn't listening, and I never realized it.

One day, a good friend and business acquaintance of mine invited me to look at something. Dustin said, "Do you have ten minutes to watch this video?" In all honesty, I didn't think I could afford to take the time given all the work I had to get done, but I said, "Sure."

I tell you what—it got my attention and gave me hope. I hadn't felt that excited for many years before and maybe ever. I thought that this was the answer for which I had been looking, and I would finally be able to get my life under control. I would have time to spend with those I love and go places with my kids, family, and friends. I believed that I would do the things I had wanted to do for many years.

I was going to be a millionaire—that was what I had told my dad and a man (who had come to sell insurance) when I was about sixteen years old. They were asking me questions for insurance purposes, and I said, "I don't need insurance— I'm going to be a millionaire by the time I turn twenty-five." They both looked at me and chuckled as if to say, "Yeah, right. That's not going to happen."

My heart sank, and my hope vanished for twenty-plus years because I chose to believe them. They had not done it themselves, nor did they know anyone else who had done it. When I saw the video Dustin had shared with me, my heart, my hope, shot up like a rocket! I felt alive again. I was going to do this. I was going to change my life. It felt like I had started a new chapter in my life's journey to fulfill my purpose.

I began to invest in myself more every day, using it as an opportunity to build a business that would provide for my family and friends. I was excited to help others achieve what they had hoped to accomplish for years but couldn't find that for which they looked. I knew I had something that would have an impact on thousands of lives and their communities to make a difference in the world. I decided that I would no longer work twelve to sixteen hours a day and seven days a week including holidays. I was going to travel and enjoy life. I invested some time into a business a few hours a week. I was making progress, had made a few ranks, had received some recognition, and I got back more than I had initially invested. It was a good start. I had never done it before, but I had a team to help me and the support of people who got no direct compensation for helping me. It was unbelievable! I will be honest: it felt good. I was energized by the ability to make a difference in my family's lives and futures. I was able to fit the business into my schedule and add another stream of income without taking away from my construction business. I thought it was great and simple—I could build the business just like I had Decor-N-More—the name of my construction business—by word of mouth. I had the right idea—at least, I thought I did, but it wasn't as easy as I thought it would be. I had started my construction business on a foundation of honesty, integrity, a strong work ethic, outstanding quality, and extreme cleanliness, and it grew by word of mouth. I became very well known for those values and fairly successful in the eyes of others. I believed I could give a repeat performance in the network marketing/direct selling industry.

When I started, it appeared that I would do just that, until I attempted to share my incredible opportunity with two of my best friends, who wouldn't even look at it. I know it wasn't their fault, as I hadn't brought it to them in the proper way. In other words, I didn't follow the system! My heart sank, and my excitement depleted. I wanted it more for them than I wanted it for myself. I felt sick and rejected. I didn't understand why, and I wasn't sure if I wanted to move forward with the business. What was going on? What happened? I no longer had the self-confidence I had with my construction business.

About a year later, a coach of mine, Scott Haug, introduced me to a program called "Thinking into Results" by Bob Proctor and Sandy Gallagher. Scott and his program, "Thinking into Results 2.0," which he had created in conjunction with "Thinking into Results," was instrumental in helping me discover the paradigm shifts in my life. I discovered the answers to many questions, some of which I had yet to ask. My identity is composed of my impressions of who I think I am, but more importantly, who I think I am not. You see, when my friends decided to not look into the business opportunity with which I had presented them, I had taken it to heart, as if they had rejected me, and my self-image went back to when I was sixteen and was laughed at because I would never be a millionaire. I had defined my identity as being a confident construction businessman. I felt secure in who I was and what I was doing. I had learned from and worked with my friend/brother, who was very knowledgeable and experienced.

Aaron (a.k.a. Harv) and I had worked together for many years. He had mentored and trained me to be knowledgeable in the field of construction, and the business was known for being reliable and dependable, even though I was not known in the direct selling industry, nor was I trusted to be experienced or knowledgeable. I did not identify with the industry as I had in the field of construction. So, when I had had a few more no's, and I was not able to handle the emotional impact of the rejection, I shut down.

My hope, my belief in who I was not, was reflected in my self-image and self-confidence was gone. I began to self-destruct and felt more miserable than I can describe. Many areas in my life started to fall apart, though I still studied the program and continued down the path of personal growth and development. I still listened to and attended coaching calls with Scott, but I did not know what I should ask to turn things back to moving in a forward direction. I also stayed connected with my business via leadership calls and events. I began taking a "Discipleship Training Institute" class through the Grace Community Fellowship—the Church I attended—led by my friend and pastor, Steve. You could honestly say that I genuinely was digging deep, and I was involved. My self-awareness was growing, and I began to take notice of what seemed to be seeking my attention—my self-image and how I pictured myself—as Bob Proctor would say, "Who I am not."

I don't consider myself religious, even though I was raised under the Church's umbrella. I have come to know and understand my creator. I have always believed in God, but I didn't follow him, nor did I have an understanding of how to apply God's word effectively in my life. I didn't identify myself with or understand how I had been created in God's image. I am not going to get into great detail—I will save that for my next book—but I began to read many of the books in the Bible regularly. With an in-depth study of God's word, John writes, "Dear friends, let us continue to love one another, for love comes from God. Anyone who loves is a child of God and knows God (1 John 4:7 NLT). In 1 John 5:1 NLT, he writes,

"Everyone who believes that Jesus is the Christ has become a child of God. And everyone who loves the father loves His children, too."

Through my reading, studying, and listening to my heart and others, I realized that my self-image was not healthy and that I believed "who I was not." I had become someone who I had not intended to be, nor did I want to be. I have always felt as though God was on the outside of me and at a distance. When I realized I was not my true self, I said, enough, and I vowed to change my way of thinking.

I began changing my self-image after I understood from where my true identity came, and with whom I had defined it. In Genesis 1:27 NLT, Moses wrote, "So God created human beings in his own image. In the image of God, he created them; male and female, he created them." In a letter from Paul to the Colossians (Colossians 1:15), he writes, "Christ is the visible image of the invisible God. He existed before anything was created and is supreme over all creation." I accepted my new identity, believing that I was a child of God. I am reminded every morning when the sun rises that I am grateful, and I will listen to who God says I am because He is the one who created me, He loves me, and He created me for greatness for His glory. My life has been transformed since I recognized that my identity had been stolen. More accurately, I had misplaced it, allowing the words of others, who they think or say I am, to define my thoughts.

I discovered what I needed to do. I had to listen to my Heavenly Father, accept and believe in who God says He is, who He is, who He says I am, and who I am becoming. Paul wrote to the Corinthians, "That is why we never give up. Though our bodies are dying, our spirits are being renewed every day" (2 Corinthians 4:16). He wrote to the Ephesians, "Instead, let the Spirit renew your thoughts and attitudes" (Ephesians 4:23). Paul wrote (Romans 12:2-3),

"Don't copy the behavior and customs of this world, but let God transform you into a new person by changing the way you think. Then you will learn to know God's will for you, which is good and pleasing and perfect. Because of the privilege and authority God has given me, I give each of you this warning: Don't think you are better than you really are. Be honest in your evaluation of yourselves, measuring yourselves by the faith God has given us."

My new and true self-image has impacted every area of my life in significant and powerful ways. A mentor of mine, Tony Robbins, said, "Human beings absolutely follow through in who they believe they are. We stay consistent with who we believe we are. We define ourselves. It becomes the glass ceiling that controls us." I agree and believe that we need to be ourselves, the true nature of which we were designed, if we are to progress and produce an abundance in life, naturally, and with the effort to raise our standards. We gain momentum through the rituals, disciplines, and intentional habits we put into practice in our daily lives. When we make values and serving others a priority in our lives, our true identities reveal themselves. Our lifestyles should reveal the light and love with which we were created and share it with the world to reveal the light in the world's darkness. I like a quote by Leland Val Van De Wall: "Let us not look back in anger, nor forward in fear, but around us in awareness." It is important to stop old conditioning or beliefs that no longer serve from preventing you from becoming everything for which you have been created, are capable of becoming, or earning all you choose to earn. Author Wynn Davis once said, "Every conflict is a set of opposing ideas. All of us have had the feeling of knowing what we should do on the one hand and doing what we feel like doing on the other. By understanding the two opposing forces warring within us, we come to a knowledge of the truth: we no longer remain slaves. A clear understanding will make us masters." We must all begin to view conflict and challenges in life as opportunities for growth in our character and spiritual lives. Make it a point to see and remember the importance of your self-image and come to know the power brought about by knowing your true identity.

BIOGRAPHY

Devon Kurz has used personal and spiritual growth for great success in the construction and remodeling industry. He is currently investing time and energy building a business through work ethic and word of mouth. He is excited to serve and help those who are working for and looking to get what they truly want in life. He is excited to lead anyone coachable to the top of their careers. Devon spent over forty years in rural Nebraska. He has two children who are now young adults. He currently resides in Oregon.

Contact Information
Facebook: https://www.facebook.com/devon.kurz

CHAPTER 11

MILK AND HONEY
By Dominika Blum

When I was three-and-a-half-years-old, in November of 1978, my mama and I were at a clinic in Warszawa, Poland, returning a strut device that had been attached to my hips and legs for the previous eighteen months. I had been delivered with forceps and a vacuum due to a complicated birth. I was born with damage to the central nervous system, head deformation, hydrocephalus, stretched sternoclavicular and mammary muscle, hypertrophic pyloric stenosis, and hip dysplasia. The hospital released me as a child from the "risk" group because there was nothing they could do for me.

Nowadays, doctors recognize such conditions and have x-rays readily available if needed. Back then, in communist Poland, one could wait up to three months to do a simple x-ray. Doctors did not want to give referrals to see specialists as they believed their own knowledge would suffice. Thank God, my mom was determined to heal her baby. We traveled all over Warsaw on a bus to find experts who would help so she could get me the aid critical to my wellbeing. Though she went through hell and back in the first few years of my life, she didn't give up. She was the one who taught me determination and resilience, and for that, I will be forever grateful.

After we returned the leg braces, we had to get my three-year-old leg muscles strong again. Mama got me a tricycle so I could ride around our tiny apartment to exercise those little legs. As my body got stronger, I went from tricycling to walking, running, and dancing like no one was watching. I sang, and I danced everywhere you could imagine. Mama signed me up for all types of dance classes at our city's Dom Kultury (House of Culture). Since then, dancing is one of my favorite things to do, and I often dance in my kitchen or anywhere, really. The feeling of great rhythm flowing through your veins is one of the best things in the world, and it's an invigorating mood changer, too. Spiritual teacher Eckhart Tolle believes, "Life is the dancer and you are the dance." I couldn't agree more.

Eleven years later, on May 3, 1989, riding in a yellow cab from O'Hare airport to Jefferson Park was a ride I will always remember. Everything was so big. The cars were long and wide. The highway had so many lanes, and the taxi driver spoke in a funny language I did not understand. Everything looked so different, so alive and fresh. The air smelled of limitless possibilities. Because of my Polish-American stepdad, we moved to Chicago, Illinois. The United States of America. The greatest country in the world. The country of milk and honey, the land of the free, where everything is possible.

I was excited about the move, curious about the unknown, and eager to embark on this new journey. I knew it would be hard, but I did not expect it to be as hard as it turned out to be. Late nights of studying became the norm, as did learning a new language, memorizing the words, the grammar, the slang, and the "in-words." Learning and understanding the United States Constitution and history to fulfill my eighth-grade requirements without being fluent in English was challenging. The Internet and online language courses did not exist at the time. There were no smartphones and no apps to teach word pronunciation. I treasured the daily disciplines my mom taught me as a child. The belief that I could learn a new language, that I could achieve greatness, grew stronger. I had no other choice. If it was meant to be, it was up to me.

The first year in my new world was one of the hardest in my life. I cried myself to sleep most nights, but to thrive and not just survive, I had to rise above it—rise above being made fun of and bullied at school, rise above my cruel school mates. Rise above it all. We have to persevere to succeed in life. One of those successes was when I earned 97% out of 100% on my US Constitution exam.

Success. /sək'ses/ Noun, the accomplishment of an aim or purpose. To succeed, one has to know exactly where one is going, have a clear vision, put in a continuous effort, believe in yourself, do the right thing, be consistent, have grit, be grateful, and, most importantly, love what you do. When you love what you do you will not work a day in your life. Success starts with daily habits and discipline. The only place success is found before "work" is in the dictionary. We can't expect to succeed if we don't put in the work.

My mentor taught me to win the day, every day. Every day do what you can control. Do not worry about or waste your time on anything else. Just focus on what you can do right now. Focus on what you can do today. Ask yourself, what can you do right now to start achieving the goals you have in your heart. Start with the end result in mind, see the prize, and pay the price to get what you want. Envision exactly where you are going, have a plan, and work that plan. Make decisions based on faith and not fear. Henry Ford once said, "Whether you think you can, or you think you can't, you're right;" do your part, and God will do His.

After high school, the question wasn't if I would be going to college, but which college I would go to. I went to one of the best universities in the nation for my major, and I graduated with honors and a Bachelor of Architecture Degree. One year later, I realized that being an architect was not what I wanted to do with the rest of my life. I wanted to change lives. I wanted to travel the world and embrace all of the life experiences I could. I did not want to be stuck in a traditional nine-to-five, and I started to look for a way out. I answered an ad in the paper with the headline, "Airline Attitude!" I had found it. I was so excited! This was my first introduction to working in sales, which was a highly competitive and flexible environment, and I loved it. I loved helping people. I loved the numbers game. I loved the pursuit. I loved the excitement of it all.

My work was based on commission pay only, and not all months were as lucrative as the one before. I started waiting tables part-time as another stream of income while I figured out the sales game. Man, oh, man—did I get an earful of OPO (other people's opinions). Note to self: never listen to anyone who does not have what you want—this applies to all aspects of life—relationships, money, health, wealth, lifestyle, education, experiences, and the like. If we buy someone's opinion, most likely we will end up buying their lifestyle. According to experts, the average toddler hears the words "no," "don't," or "can't" four hundred times a day on average. It's no wonder we don't always believe in our own abilities, as we are trained our whole life not to.

I was twenty-five-years-old, and for the first time in my life, I was introduced to the idea of personal development. "What do you mean we have to work on our minds?" I asked my mentor. The concept was so foreign to me. I then realized that I already had everything I needed to succeed. I had the control to make my life what I wanted it to be, and it all started with what I fed my mind. I began to read good books like *Think and Grow Rich* by Napoleon Hill, *The Magic of Thinking Big* by David Joseph Schwartz, *The Power of Positive Thinking* by Norman Vincent Peale, and *The Power of Your Subconscious Mind* by Dr. Joseph Murphy, to name a few. I nourished my mind with positivity through good books, CDs, workshops, seminars, affirmations, positive self-talk—whatever I could get my hands on—every single day. Our brains are like gardens—what we feed our minds matter. What we focus on expands. Ask yourself, are you feeding the flowers or the weeds in your mind's garden.

Over the last twenty years, I have invested a lot of time, energy, and money in my personal growth. The best investment one can make is in one's self, after all. We all need to work on ourselves every day. Just like taking a shower or a bath, we need it daily. Work on your mind muscles daily. This is a lifelong process. The minute we think we know everything there is to know, we go right back to where we started. Be intentional with who you include in your circle. We have the choice to surround ourselves with positive, loving, driven, goal-oriented people who will uplift us and push us to be and do better, believe in us, and tell us the truth. We must seek out action-takers, powerhouses, go-getters, leaders, achievers, doers, kind, selfless people with huge hearts. I choose to learn and broaden my horizons on a daily basis, using that knowledge in my life. I don't believe that knowledge is power; I believe that knowledge combined with action is power. We can be the most educated people in the world, but if we do not do anything with that knowledge, it does not matter how much we know. To get what we want in life, we must ask for what we want and not what we don't want. Most importantly, what we want must be backed up with love and action.

Ralph Waldo Emerson once said, "The only person you are destined to become is the person you decide to be." Every experience makes us stronger and wiser and drives us to make the right choices. 2012 was a year of major life decisions for me. I resolved to cut myself off from friendships, relationships, habits, and behaviors that did not serve me positively, and I found my purpose. I love exploring and traveling the world. My mom is my biggest inspiration in this area, as she's the one who got me hooked on traveling and discovering the world when I was little. Travel humbles us by showing us how little space we occupy in the world. It teaches us in the most profound ways, stretching our way of thinking, our beliefs, and our imaginations. Saint Augustine of Hippo described this idea when he said, "The world is a book and those who do not travel read only a page." Creating memories with friends and loved ones across God's earth has an impact that will live in us forever. I have traveled to forty countries and thirty US states so far, creating a life from which I do not need a vacation. I am committed to reading the whole book and taking a bunch of people with me along the way.

One must serve others from the heart and see the world from a place of love if one is to be successful and feel fulfilled. The more we love, the more we get love back. Love is a purpose. It is our true power. It is what inspires us and what changes us. Love is at the core of success. Love is the key to all. "We love because He first loved us," 1 John 4:19. God first loved us, so love yourself just the way you are. Love one another, speak life, not death, into your daily life and trust God to do His work. He has a grand plan for all of us. A part of God's plan for me was to serve in Guatemala. I volunteered at San Martin municipal for the first time in 2015 with the Hug It Forward organization that helps build schools from plastic bottles filled with trash. I immediately fell in love with the culture, the beautiful people, children and their love for one another. The El Chocolate community stole my heart and changed my life forever with their big hearts and warm smiles. I returned to Guatemala three more times and led volunteers from across the world to build bottle schools. I am stoked to go back this year for the fifth time in my life, this time with college students from a local university. I look forward to watching them grow into servant-leaders on this volunteer voyage.

Our attitude and mindset, showing up, being present in the moment, and having the belief that you will win are all a part of victory's bigger picture. My good friend, Roscoe, says that you've got to show up and participate in your own rescue. We all live in a time of abundance, but when we make decisions based on fear, we believe the lie that we live in scarcity rather than abundance. Whatever we believe becomes our truth, turning our perceptions into our realities. Our words are powerful. We speak life or death into our lives daily by what comes out of our mouths. We either build ourselves up or tear ourselves down. It is so easy to be negative, to complain and play the victim. I challenge you to change your way of thinking over the next thirty days and start paying attention to your thoughts and words. When negative thoughts manifest, change them to positive ones. Love instead of hate. Think before speaking. Compliment instead of complaining. Be the victor God made you to be. There is more than enough love, abundance, prosperity, wealth, and goodness in this world for everyone. God gave us dreams and goals for a reason. He sent us into this world not to be in it, but to transform it.

My purpose and mission are to help millions live their best lives. By aligning our actions with our intentions, we can make our dreams come true. It does not matter where we come from. It matters where we are going. Begin by writing down your goals and affirmations and read them daily. Visualize what you want to do and be. Start from a place of love. Believe in yourself, practice positive self-talk and take massive action. Relentlessly persevere daily, no matter the circumstance. Though it sounds cliché, don't sweat the small stuff. Do your part, and God will do His. Dream BIG and go after it. Get your milk and honey, friend. You are worth it.

BIOGRAPHY

Dominika Blum is a sought-after leader with a vibrant and contagious attitude. Best known for her laughter and servant leadership. She is a passionate entrepreneur, speaker, and international advocate for education. Dominika has been inspiring people to live their best lives for the last twenty years. Her mission is to transform millions of people by helping them rise to their highest potential. She believes everyone with the right attitude and mentorship can achieve their goals. Dominika is a life-lover, travel fanatic, dream-pursuer, and relationship-builder. When she's not traveling, you can find her encouraging others, learning, and reading. Her guilty pleasure is ice cream and romantic comedies. She is married to the love of her life, John, the best husband ever. They live happily in South Dakota, USA.

Contact Information
Facebook: https://www.facebook.com/sunattitude
Facebook blog: https://www.facebook.com/UwierzwSiebieDominikazWarszawy
Instagram: https://www.instagram.com/sunattitude
Twitter: https://twitter.com/sunattitude
LinkedIn: https://www.linkedin.com/in/dominika-blum-53046a19
Email: dominikablum@gmail.com

CHAPTER 12

THE MAGIC OF LIVING OUTSIDE YOUR COMFORT ZONE

By Helen Kithinji

I am backstage, surrounded by media production machines and the production team. In the background, I can hear loud hip hop music that sounds like a million people singing along. The auditorium is electric. This extravaganza is making me even more nervous, just as one of the media guys is fixing my mic to go on stage. This is my first time ever hosting an international event. This one is a three-day business training event in Johannesburg, with over 4,000 entrepreneurs from all over Africa in attendance. After my mic is fixed, I ask the production director to give me a couple of minutes to calm my nerves. He didn't need to know that I was going to pray and then telephone my coach. My heart was racing, my palms sweating, all of which was reminding me of a past experience in a similar situation.

I was assigned the role of introducing my singing group during one of our performances at the Conservatoire in Nairobi, back in my high school days. Our music coach Mr. Walshaw had prepped me well. Once on stage, I was to take one step forward from the other 11 singers, introduce the group, then give an outline of the classical pieces that we were going to perform. I'd spent two weeks memorizing my short speech and practicing every step, the pose, and the smile. I had mixed feelings, as we waited for our performance slot. I was feeling excited about the opportunity, yet nervous and scared at the same time. When our time came, we walked onto the stage, formed a neat semi-circle, while the audience applauded. When the applause was over, our music coach signaled me to start. I stepped forward to do the introduction. I could barely see the faces of the audience through the dim lights, but I could feel their eyes on me. The pin-drop silence paralyzed my mind and my mouth. I went blank. I couldn't remember what I was supposed to do or say. I stood there tongue-tied, trembling, staring into the audience like a statue. That one minute felt like five minutes and only ended when Mr. Warshal snapped his fingers from the far end of the stage. Everything streamed back into my mind, and, judging by the applause after my speech, I had done a good job, after all.

The memory of this experience gave me a tremendous fear of public speaking for a long time afterwards. Indeed, when I was working as a public relations manager in the bank where I worked many years later, history repeated itself during a bank-customer relationship event. My boss missed his flight, and I had to host the event. When I stood on stage to kick off the event, my mind went blank again. I remember the sweaty palms, my heart racing, and my whole body trembling. After a minute of paralysis, I gathered my courage and managed to run the day's program smoothly. This was the last incidence of my stage paralysis, but definitely not the last of my stage fright. Over the years, I have had numerous occasions to speak onstage, every time experiencing nervousness and anxiety, especially in the minutes just before going onstage. Today, I no longer have those mind-locks and am more relaxed on stage. Taking training courses on how to be an effective speaker and practicing every time I speak means that I get better each time.

You would think that having spoken onstage many times before, I would be confident about hosting my first international event. After all, the only difference here was the size of the stage, the audience, and the auditorium. I had prepared well, carefully choosing my outfits, accessories, and my hairdo, and I'd practiced my opening speech, researched some stage jokes, and watched videos about emceeing. But here I was, deathly nervous, almost about to back out at the last minute. But my determination to be a highly sought-after speaker got the better of me. I spoke to my coach, who guided me through a centering process. I took some deep breaths, whispered a prayer, and was finally ready to face my fear. As my name was announced, I walked onto the stage, raised my hands up in the air, inviting the audience to stand and dance with me. As I danced across the stage, my stage fright was suppressed. My courage was boosted by the love and energy of the audience. The audience danced along happily, with some people shouting my name and sending flying kisses, fueling my courage and strength. I had interacted with most of the people in the audience in person or through social media. Most had been following my journey, as I built my global travel business, becoming a top thought-leader, and earning a million dollars in commissions over six years in our part-time, home-based business. It is my success in this business that earned me the coveted chance to host this event. My journey to earning a million dollars is one I can summarize in one sentence: "I found my fortune outside my comfort zone." I've found that my growth always comes from pushing myself through uncomfortable experiences.

When I look back at my early years growing up and then as a young professional, I've found that my major achievements were always preceded by intense fear, anxiety, and the temptation to quit. As human beings, we're inclined to try to protect ourselves from the unfamiliar. We tend to resist that which pushes us out of our comfort zone and toward change. Yet it's through change that we grow.

All personal breakthroughs begin with a change in beliefs. That's what I've learned over the years; to make a breakthrough in our personal development, we have to start by challenging our long-held beliefs that were formed during our upbringing and our life experiences. During my early years, my beliefs about success were shaped by my parents and the community around us. For instance, my father was a traditionalist who believed that one only needed to work hard in school and get good grades to secure the ideal—a job with the government. The most coveted government jobs at the time were high-school teacher and district administrative officer because they were seen to provide job security and a modest salary to live on until retirement at sixty.

My observations told me that the majority of civil servants lived peaceful, average, event-free lives, providing the basics for their families, but never really achieving massive success. Because I had bigger dreams for my life, I decided to focus on securing a job with a multinational company or a bank. My father and his generation grew up with the belief that working for the government presented the best opportunities and that it was difficult, almost impossible, to qualify for a job in a private enterprise. After making applications to fifty-three banks listed in the yellow pages of an old directory, I secured a position with the largest bank in the region. Fifteen years later, I exited the banking industry as a highly successful senior bank executive to become an entrepreneur.

Ten years have passed since then, and I have grown into a successful entrepreneur. I run a global business, have invested in a couple of other businesses, including a financial institution, where I serve on the board. I've achieved accreditation for executive coaching and often publicly speak to and train entrepreneurs. I'd like to share the key principles and lessons that I've employed to create a system for continuous growth in all areas of my life, including my health and fitness, parenting, relationships, career, business, lifestyle, and spirituality. Most recently, I achieved the milestone of having earned a million dollars in my home-based business in only five years, an achievement that I never thought was possible.

1. **Identifying and replacing limiting beliefs**

    Personal development and my coaching training exposed me to the power of questioning every limiting belief that stands in my way of achieving my goals. Psychologists have proved that our brains are always trying to move us away from anything unfamiliar or that we think might cause us pain. By asking powerful questions to interrogate such beliefs, you can overcome and replace them with supporting ones. Some of the questions I ask are:

    "How has this belief served me in the past?"
    "How has it limited me in the past?"
    "What will it cost me in the future if I don't act against this belief?"
    "What more empowering belief can I replace it with?"

It's time to get rid of beliefs that do not serve you! Make a list right now, ask the questions, and then commit to replacing them with empowering ones.

You'll no doubt find that limitations will also come in the form of justifiable reasons or excuses. These, too, must be subjected to the same process outlined above. For example, in the past, I told myself that I couldn't go to the gym because I didn't have time. After going through the questioning process, I realized that neglecting my fitness would cost me later in life in the form of medical bills. I dropped the excuses and hit the gym.

2. **Setting high standards**

   To create intense desire and drive to achieve audacious goals, I set high standards for myself. During my banking days, my boss was always astonished at the aggressive targets that I set for myself. This practice saw me ascend in my career faster than my peers. I continue to use this principle in my businesses and personal projects.

3. **Commitment to continuous learning and personal growth**

   Your ability to expand your mind and commit to lifelong learning is critical to your success. My commitment has seen me get ahead in every aspect of my life. There are three forms of learning, namely; *Maintenance Learning* aimed at keeping you in your current position; *Growth Learning* to increase your skills and knowledge, and lastly, *Transformational Learning,* which contradicts and challenges your beliefs, triggers new insights, and creativity to move you forward beyond that which you believed was possible. You can continue the learning journey through certified training events, online courses, books, and validated content.

   All successful people that I know of commit to continuous learning, with most reading at least a book every month. Consider the extreme reading habits of billionaire Warren Buffet, who is reported to read fifty books a year and spends hours reading hundreds of pages of business reports every day. Could this routine be one of the keys to his massive success?

In Japan, they have a continuous improvement system called Kaizen. So, based on my experience and knowledge, I recommend committing yourself to the habit of constant and never-ending learning for growth. This is an investment in yourself that will enrich and enhance your relationships, families, businesses, and communities.

4. **Working with a mentor**

    Finding the right mentor to work with can propel you on your way to success, allowing you to be guided by someone with superior experience and knowledge of your chosen field. It's an effective way of shortening the learning curve and avoiding unnecessary mistakes. Your mentor will help you to challenge any beliefs you hold that may be blocking your way to success and boost your self-confidence, leading to further breakthroughs. Throughout my life, my mentors have played an important part in supporting and cheering me. They've celebrated my achievements when I've been successful; and supported me through tough times, helping me to understand my limiting beliefs and pushing me to overcome them.

5. **Discipline yourself to take action consistently and persistently**

    One of my mentors has always said that success doesn't come easily or quickly, you need to be patient with the results but impatient with taking the right action to take you forward.

My message is: Consciously and consistently push yourself outside your comfort zone in order to continue a positive progression. Learn to notice when your beliefs are limiting you and replace them with empowering ones. Imagine that they are the brick wall that you must crash through to achieve your dreams. Don't let the fear of failure stop you. Overcome the fear by doing what you fear anyway. If you're willing to do what is uncomfortable for you to move yourself in the direction of your ambitions and have the courage to tackle the unfamiliar, even when you have no guarantees for success, you will ultimately succeed. It's part of the process to allow yourself to learn through failure, and trust that every time you fall, you'll get up having learned a lesson that puts you closer to achieving your goals. As Thomas J. Watson, former CEO of IBM, put it, the formula for success is simple— double your rate of failure.

Living outside your comfort zone will bring you heightened levels of energy, passion, and an enduring state of happiness and satisfaction. It is simply MAGICAL!

BIOGRAPHY

Helen Kithinji is a successful entrepreneur, investor, an Accredited Executive Coach, and author of one of the most powerful personal development books published: *Profiles on Success with Helen Kithinji*, which also features other global best-selling authors. An MBA graduate, her story about growing to executive management in her fifteen-year banking career, becoming an entrepreneur, and establishing a multi-million-dollar global business in less than five years continues to provide inspiration and encouragement to others in pursuit of their dreams. She has been featured in leading business publications, including the Voyager Magazine, Business Daily, The Daily Nation, Business for Home, and The Standard newspaper. She has won several awards, her latest recognition being the achievement of a million dollars in sales commissions in her online business. She also enjoys a rich, happy family life with her husband—Ronnie—and their two adorable sons.

Contact Information
Facebook: https://www.facebook.com/rahalinks
Instagram: https://www.instagram.com/helenkithinji/
Website: https://www.helenkithinji.com/

CHAPTER 13

I Choose Me

By Ilioara Ormenisan

Why do human beings give up on things so easily?

Has this question ever come to your mind?

My entire life, I've been seeing people around me give up easily on their dreams and goals. Since I realized that, I've been wondering why that is, why don't they want more for themselves. I can't be the only crazy person wishing for things that may seem unachievable.

I've been surrounded by people saying that they want more money and fulfillment in their lives. Yet, when opportunities come, they seem to be blind to them. It's almost as if they're putting on a blindfold and embracing NO with ease, instead of making more positive choices.

Looking from the outside, we can see that people are usually capable of doing what they want, but they just don't see it.

A while back, I was one of those people asking what my purpose in life is. Sometimes, I'd get frustrated, as I bought into other people's ideas about what life should be. Back then, when I wondered about this, I thought of all the many things I know how to do, like baking, sewing, decorating, handmaking crafts, cleaning, ironing, nail technician, etc. But none of that felt like my true mission in life.

I found myself making negative judgments about myself and thinking in the same downbeat way. I felt similar to the way I did before my husband and I started our relationship. We've known each other for 26 years, but our romantic relationship only started seven years ago. Before we started hanging out, I was single for two-and-a-half-years, and every date I went on was a failure. Plus, all the time, I had the feeling that everything was falling apart. At one point, when I thought about this, I remember thinking that I was sure what was "mine" was waiting for me, but wondering when I was going to find it. I often felt the same about my purpose in life, too.

After all this destructive self-talk, I realized that I wasn't heading in the right direction and needed to start looking at things from a different perspective. What if, for instance, my mission in life was to be happy, spread happiness and joy around me? That could lead to amazing situations. I started to enjoy the little things in life more and began to see the good in people, even the ones who annoyed me. But the most important change I made was to start being more grateful for everything around me—and I mean, absolutely everything.

Have you ever been in a situation where you felt it was right to trust your gut about something and, despite anyone else's arguments, you went ahead and did it anyway?

I surprised myself by acting this way even when my family and friends disagreed with my intentions. But I knew I had to trust myself and take that leap of faith in order to move forward. It wasn't always easy. I have been blessed with challenges, but my intuition has usually proved that I've made the right decision. You might wonder why I say I have been *blessed* with *challenges?* Let me explain: By deciding to change my outlook, I came to the conclusion that every challenge (I stopped calling them problems long ago) I faced in my life could be overcome if I faced it with courage.

I believe this has helped me become a better and stronger person today. I hope you understand the message I'm trying to get across—that challenges are actually *blessings* in disguise. It's not so much about the challenges that we encounter throughout life. It's about how we face them and deal with them.

I am 100 percent certain that every single one of you, at some point in your lives, have come across a situation which, at the time, left you breathless or completely stressed out. Then, after tackling the issue, you looked back and thought that it wasn't really that bad after all. I'm not sure how true this statistic really is, but it certainly feels as though one percent of life is what happens to you and 99% is how you react to it.

Please allow me to share my story with you: I always considered myself an ambitious person, but kind and sweet at the same time. Although I saw myself as a fighter in life, committed to succeeding, and working hard to achieve my goals, I still gave up on things from time to time, which led me to the question we started out with. For a long time, I couldn't find an answer to that question of why we sometimes quit things we've started. Maybe I wasn't thinking about it hard enough? Besides, I could always blame it on human nature—that this is who we are as people, and that we are capable of changing our decisions overnight. Have you ever felt this way, or is it really only me? I don't think so.

People might call me crazy and a dreamer, but I believe in magic, and no one in this life or the next can shake my opinion on that. I always considered myself as having a special something, but never really knew what it was. But now I feel that I do.

Do you know that feeling when people are trying to force you into choosing between two different options? That's not me. I'm not the type of person who makes those kinds of choices. That's not my reality. I'm talking about choices that have a long-term effect on you, like quitting college, a relationship, or giving up on a dream that others say is crazy. Isn't it better not to buy into their points of view, but do what you feel is right? I remember a situation that happened before I went to university: I was in a dead-end situation but couldn't accept it. I had a choice between going to university or staying in a relationship. But I didn't want to choose—I wanted both. What happened was that, by not choosing, as most people would, the challenge of having to find ways to overcome obstacles to having both sent my brain to work. And it came up with solutions that worked. This just proves my point that, when you ask yourself questions, your mind is forced to answer.

I'd like to share another story from a few years ago about this point of view: Three months after I got married, I took the decision to leave my family, my home country, and move abroad without my husband. The reason I did that was because I wanted us to own our own home. To achieve that, we needed money for the deposit. So, I emigrated to the UK to find a job and save up the money for it. If I'd let my family's arguments against going influence myself, I would have stayed at home. But I was determined to follow my instincts that it was the right thing to do, and I would succeed in my aim. Sometimes, thinking logically about something just gets in your way. I thought about the fact that I was going into the unknown, I hadn't even got a job in the UK, but I dismissed any worries.

Instead, I listened to my heart, which told me to proceed with my plan and follow my dream. Now, it sounds so easy when I write about it, but at the time, I remember having days when I felt like the sky was falling in. I had it planned that I'd work for six months, save all the money I could, then go home. Except that, in the beginning, I was told that we needed 5% of the total mortgage cost, but when my husband checked six months later, the realtor suddenly wanted 20% and gave us a one-week deadline to find it or lose the house. Once again, I set my brain to work and looked for solutions to the challenge. This is another example of the approach I take with all challenges. I never accept NO for an answer when it's something I really care about.

The important point I'm making here is that I knew what I wanted and fully committed myself to achieve it. I chose to believe in myself and my gut feeling that my choice was right.

I've realized that when I choose to commit to my deepest and, some might say, unacceptable desires, dreams, or goals, which I believe in with every molecule of my body, other people's opinions can't sway me. When you're committed to yourself, you will never give up doing what you think is right, because that conviction comes from inside you. It comes from the heart, and no one can stop that burning desire. Do you remember hearing stories about mothers saving their children from death by almost superhuman effort? From somewhere, they find the strength to save their child, even if it means lifting a car or breaking thick window glass. The same thing happens with us when we're functioning in that space where we feel our dream, our purpose coursing through every single part of our body. It's your life that we are talking about. It's not something to be treated as a New Year's resolution that you give up on after a few days, weeks, or months.

When you truly decide to commit to yourself, when you take charge of your own life, then it's not important how many followers you might have, it's about knowing where you're heading, following your gut instinct, going with your intuition. You have to trust, have faith, and certitude that, when the time is right, you will succeed in your goals because you know your purpose.

Throughout our lives, we let ourselves be influenced by so many people, by family members, by doctrines, but who's to say that they're right and you're wrong? We tend to be judgmental about our actions and often feel guilty about our decisions. We live life doubting if we've made the right choice. But, what if you can choose to create new and different possibilities? Please, would you do me a huge favor? Could you stop living other people's lives and start living your own, in the way your soul truly desires? Don't let yourself be overwhelmed by all this, just take the reins of your life, even if it means breaking out of your comfort zone.

Right now, you might be thinking, *"Pffi! I already live the life I want."* But is that true?

If that's you, then you're not lying to me, you're lying to yourself. Let me give you a tip about choosing—did you know that you can choose to change your reality every ten seconds?

I encourage you from the bottom of my heart to take a few minutes for yourself and take a retrospective look at your life; think about the things from the future you can bring to the present; ask yourself how you really want to live. You can even write down a list of the changes you desire to make in your life. Now is the time to erase all negativity and stop postponing things: It's time to CHOOSE YOU and take control of your life.

BIOGRAPHY

Ilioara Ormenisan is an ambitious, caring, and hardworking entrepreneur. She is creative, persevering, and passionate about traveling. She has a strong desire to make an impact on the world. Her mission in life is to spread happiness, hope, and joy around the world by encouraging others to follow their heart's desires.

Contact Information
Facebook: https://www.facebook.com/ilioaraormenisan
Instagram: https://www.instagram.com/ilioara/

CHAPTER 14

Seven Living Generations

By Ilze Strauta

At present, seven living generations span across the planet. These generations are seven distinct groups of people born within a defined period who shared similar cultural traits, values, preferences as they aged, and therefore have similar ideals.

Different countries have different generational definitions based on major cultural, political, and economic influences; however, here are general cultural generations.

1910–1925: The Greatest Generation
1926–1945: Traditionalists or the Silent Generation
1946–1964: **Baby Boomers**
1965–1979: **Generation X**
1980–1995: **Millennials or Generation Y**
1996–2009: Generation Z
2010–Present: New Silent Generation or Generation A

Though it is easy to fall into overly generalized stereotypes when talking about generational differences but each generation's future is affected by its childhood, the outcomes of different childhoods are similar to the experiences and outcomes of history and the future. Every generation has a set of needs, values, and dreams. However, historically each generation holds some bias against another.

People in different countries were influenced by different events, politics, and social life. Due to rapid technological evolution in some countries, people could accomplish and cultivate culture and social life faster than in other countries. It is all the more crucial to consider the characteristic differences of all generations, as today at least four to five working generations are part of the labor market, working together under one roof. This century faces an unprecedented challenge, that is, to prevent misunderstandings between generations, which could lead to unnecessary conflicts. Sooner we understand what drives different generations, sooner we find a way to interact with each other.

How do we find a common language?
Pay attention to the signs of conflicts, controversies, and needless blowouts.

The Greatest Generation

Shortly after World War I, this generation witnessed a new era—stuck between childhood and reality of the war—called Depression Era. The war led to the death of millions of people, destroyed long-established peace and economic stability, and created conflicts between many countries in Europe, Middle East, Russia, Asia, and Australia. This generation became compulsive savers, careful with spending money, and felt responsible to leave a legacy. They grew up to be patriotic and authority-abiding citizens.

The greatest generation lived without the aid of technological advances, like airplanes, radio, TV, refrigerators, electricity, and air conditioning. The concept of "retirement" was alien to them; you worked until you are dead or cannot work anymore. Divorce was a taboo, and relationships were forged for eternity.

This generation experienced not only technological and medicinal change but also a huge culture change.

Traditionalists

Postwar, this generation was surrounded by darkness, doubts, and everlasting concern, but it was at the helm of new work and business opportunities.

Peace, jobs, suburbs, television, rock 'n roll, cars, Playboy magazine, and the Korean and Vietnam War—were few of the defining moments for this generation. Women stayed at home, looked after children, and waited for their husbands to come back from work with a hot dinner. While men worked for one employer their whole lives and provided security and comfort for their families, this generational believed family to be a sacred and eternal entity (meant to be together forever), while children born out of wedlock were not accepted by society.

This generation earned and learned to save money. They planned for retirement, to make up for the hard life after the war. They have been financially prudent, the richest, most free-spending retirees in history.

Baby Boomers

Born after the war, baby boomers are part of the generation when the attitudes, behaviors, and society, in general, were rapidly changing. They were born to work, to be employees, and to be careful and patient. They are very loyal and reliable.

In Europe, after World Warr II, the society experienced a postwar economic rise. In Russia, people underwent various side effects of socialism and collectivization. The United Kingdom faced an economic crisis. In the United States, while many people served in the military, the country's economy was still in better shape. With the advent of industrialization, manual production shifted to machine production, and manufacturing companies provided jobs. Baby boomers worked to secure their children's lives.

Self-righteous and self-centered—too busy for neighborly duties yet strong desires to reset and change the common values for the good of all—this generation became workaholic. Baby boomers are liberal and modest; they were not a part of the labor market but were loyal to one employer their whole lives. They are well-suited to be a team player and oriented toward collaborations. Working with one employer your whole life was considered to be honorable.

Baby Boomers' spending habits and consumerism fuelled world economies. They became famous for spending all the money they earned. Spend now, worry later, buy it now, and use credit— they are not adept with finances. After the war, several planes were converted into passenger planes, which enabled Baby boomers to travel and experience and see more than their parents ever did.

Generation X

Acquiring education and climbing the career ladder are important for Generation X. They look for opportunities, are result-oriented, and focus on self-development and independence in their private lives and financial matters. Not only are they profit-oriented, but they also love to work and live for themselves. In professional spheres, Gen X is motivated by recognition, trust, career advancements, and financial rewards. Followed by baby boomers, they learned to find a balance between theory and practice. They witnessed the fall of the Berlin Wall, the collapse of the Soviet Union, spent most Saturday nights at a disco, MTV, microwave ovens, color televisions, video games, and personal computers.

Often called "lackey kids," both their parents worked. Consequently, this generation is typically independent and does not thrive on micromanagement, and can even be considered entrepreneurial. They focus on saving the neighborhood, instead of the world. They are tech pioneers because they saw the transition from the industrial to the digital age and had personal computers, microwave ovens, and video games at their disposal.

Gen X is not the core advocate of the government. They suffered from the increasing drug problems in schools, mindlessly used credit cards, craved high status in the society, and blindly chased money.

This group desires to achieve a higher quality of life and prefers lifelong learning and earning opportunities.

Generation Y

This generation experienced the 9/11 terrorist attack, internet era, cable television, satellite radio, 2008 financial crisis, the extreme effects of climate change, Google, Skype, PayPal, Facebook, and so on.

Pursuing passion and following interest in work took precedence for this generation. They accept difficult tasks as a challenge and are not too loyal to the employer. They want to work for themselves, and they regularly change jobs to gain financial advancements. They prefer to work in a team, expects leisure time, and can be confident yet aggressive. Growing up at the beginning of the technological globalization, they know everything about everything.

They are ambitious and have a flexible work schedule and an interesting life. They expect regular feedback from their managers, and if they feel unappreciated, they don't shy away from changing jobs.

They grew up as skeptics, but they are energetic, go-getters, and have leadership instincts with built-in self-confidence. With the technological advantage and great digital skills, Generation Y refrains from participating in a "rat race" but works hard like all previous generations. Work-life balance is important for this generation as most of them came from unhappy, divorced families, and saw their parents working two jobs.

Generation Z

Generation Z seeks freedom and self-awareness. They believe in the joys of work and leading a happy life. They carve their lives around the 24/7 principle. They expect challenge and trust at work; are realistic and individualistic. Workplace safety is important for this generation. They are self-starters, confident, and value-oriented. They are known to find new ways to make money and gain fame and profits rather efficiently. They are fascinated by dynamism and are prime multitaskers.

YouTube has come to aid for this generation. They are realistic and understand that no employee can become a millionaire. They learned early in life that time is precious. They are not afraid to take responsibilities or be in leading positions, but they separate private and work life. Gen Z is uncompromising when it comes to working on weekends.

At present, Generation Z occupies a small part of the labor market, but they are very technologically oriented, fast learners, and adapt quickly. They are accurate and creative and eager to be financially stable. They harvest information off the internet.

Their lives are built around and blended in smart devices, modern technology, internet, social media, Facebook, WhatsApp, etc. They are adept in online business techniques, like email marketing. However, their self-worth is often based on the "likes" received on social media, which makes them sensitive to criticism.

However, they appreciate working with experienced colleagues. As most of them still are students, and only the future will show what to expect from Generation Z.

Generation A

Generation A or "digital babies" are exposed to advanced technology from birth. They are born in a fast-changing world with an abundance of information. They are technologically literate from birth and would be experienced in the use of touch-sensitive or artificial intelligence-enabled devices.

Although some parents might be against allowing their children to use technology or limit its usage, such restrictions are of little benefit. Most schools are rushing to transform their teaching methods and incorporate digital technology. It might seem like a clip from a dystopian movie, but this world is the new reality where the new generation is dependent on technology.

Perhaps in the future, Generation A will realize the power of human interaction and the need for emotional attachment.

Bottom Line

Age is just a number, and technology advancements are part of our lives. I am looking forward to more development leading to more exciting opportunities.

If we shut ourselves to new possibilities and don't want to understand and accept changes, the conflict will be knocking on our doors. Had we followed "New way is not a right way," or "Nobody has done it before," humanity would not have progressed. The invention of the light bulb was not a touch of the devil, but the persistence of mankind toward a major advancement of the civilization.

The amalgamation of the work of all generations has led us here. Everyone's work is meaningful, and for that, we must be grateful to each other. Everybody expects a decent salary; a leader who listens and respects their point of view; support from others; and to lead a stress-free life. This is the basic expectation, not a stereotype.

Every generation brings in this world their own set of values and views, motivations, and attitudes. Different people and generations make life interesting. So, do not let life escape from you.

BIOGRAPHY

Ilze Strauta has studied psychology, marketing, entrepreneurship, restaurant management, art, and history of art in Latvia and Russia. She has learned, mastered, and practiced network marketing, social media and digital marketing, and business in the United Kingdom, Europe, and the USA. Strauta worked in hospitality for twenty-four years, before venturing into writing. She has experimented with several occupations: secretary, restaurant manager, event manager, human resource manager, brand host manager, and regional training manager. She considers mentoring and coaching people, who struggle to step up, her favorite passion. Strauta loves to travel, explore different cultures, local cuisines, and meet interesting people. She believes that her Latvian sense of humor has helped her to overcome the challenges of life gracefully.

Contact Information
Facebook: https://www.facebook.com/life.travel.success
Instagram: https://www.instagram.com/life.travel.success/

CHAPTER 15

ACHIEVE WHAT YOU WANT

By Iréne Wrigstedt

My story is my legacy to my children and grandchildren. I hope it will give them the courage to believe in themselves, follow their dreams, and achieve what they want in life.

I want to encourage people who face challenges and think that they lack the ability to dream big.

Everything is possible. Don't let negative thoughts limit you and don't listen to negative people. Determine what you want, express it, and embrace it courageously when it shows up. Go with your gut; it knows what's right for you.

I always have dreams. I am not afraid to fail. The goal can always be amended.

It's not that I don't listen to other people—I used to do it a lot and followed what they said even if I didn't agree. I stayed quiet to avoid conflicts.

I grew up as a lonely child with loving parents. I was their everything, which turned out to be a significant burden. Everything should look good. My father once told me I was *a good average person*. It has followed me to date. I developed a self-image of not being good or intelligent enough. I imbibed it and lived the first part of my life in the shadow of that perception.

Even though it felt lonely to grow up without siblings but not having anyone to lean on or get help from has made me strong.

Due to the experiences growing up, I made it a priority to be kind, loving, and caring toward others. I try to support and help my children and friends as much as I can.

I value honesty, love, trustworthiness, and ethics. As long as you are brave and have faith, success will follow.

I always did things outside the box even before I knew the expression existed. People often laughed at me and criticized me. I think it says more about their insecurity than about me.

Money and creativity drives me. Growing up, I experienced a lack of money; my parents were open with me about our financial condition, which made me anxious. Later in my life, being totally broke taught me the value of money. It became my goal to build financial security for my children and me.

Creativity is also an important factor. I love nurturing ideas and developing them into businesses and help people grow.

When I was twenty, I moved to Lausanne to work. My parents repeatedly warned me that it would not be possible to find a job and a place to live. But I was determined. I had been studying there after the gymnasium and had a great time studying, partying, and skiing. I moved, got a job, a flat, and continued the fun life.

After two years, I was ready to move back home. I wanted to buy a horse. Ever since I was little, I was horseback riding, so it was a long-due wish. Due to the cost involved, my parents did not support me, but I had saved enough to buy a horse.

Even though they did not support me initially, my parents did not shy away from boasting about me coping so well in Switzerland and back home my participation in horseback riding in front of their friends. That did hurt me as they did not support me from the beginning.

I was raised as a good family girl, always conditioned to do the right thing. I was trained in how to *behave* in various situations. I did not understand why it was important and of what use. Despite all this, I did my own things.

All this amounted to making me feel prestigeless, which I see both as good and bad. Good because I don't care about prestigious things, and I am not afraid of being embarrassed. Bad in a sense that people might perceive me as having no courage or impact.

Being prestigeless does not mean that I don't own prestigeful things such as nice clothes, nice cars, and a beautiful house. I have decided to indulge in these luxuries for my pleasure and happiness, not to gain validation from others. I am not boastful, just grateful though I have experienced jealousy, which came as a surprise to me.

I have never been jealous. I take full responsibility for my actions and for what I have achieved, and I am happy for other people achieving success.

I was asked to go in therapy by my husband and my business partner to become more easy to manage. So I did.

I also did some massive life coaching at Landmark—making up with your past and creating your future. It changed me as I gained insights about myself and was given tools to make changes if I wanted.

I asked for a divorce from both my business partner and my husband.

After moving back from Switzerland, I started working as a secretary to the CEO, something my parents had encouraged me to do. I served coffee, took shorthand, typed out letters, and behaved kindly. I performed well, but my bosses were still unhappy as they could not *master* me. So, I took a degree in marketing and started a new career.

In 1992, I saw an opportunity to set up a business based on a concept that did not yet exist in Sweden. The timing was right, and the market was there; I hit the ground running. I also brought in a business partner. After a few years, I expanded the business to Denmark, Norway, and Finland. I had seventeen offices in four countries with around three-thousand people on the payroll.

As I was still living with the self-image of *a good average person*, I did not assert complete authority in my business. I relied on my business partner too much, who prohibited me from talking to our staff because I did not do it the right way, and they were afraid of me. I believe that it was my business partner who was afraid of my success.

While in the business, I gave too much authority to the senior managers, which resulted in me losing control of the company.

At a turnover of a hundred million Swedish crowns, I filed for bankruptcy in four countries due to the lack of control. I also found out that people whom I trusted had cheated me.

I ended up broke. I had built a big company, helped my clients earn a lot of money, and given my employees good salaries as I appreciated their work. Everything fell apart in a few days, and no one had anything good to say about me. I was left with absolutely nothing.

As my clients were well-known, the media in four countries started to chase me to dig scandals and sell stories about me. There was nothing to find, but they were dealing with a naive person with a lack of control, so they started to concoct stories. It was an awful experience.

Despite the downfall, the business idea was still fresh, and the market was still responsive. A businessman saw this as an opportunity and stepped in to earn money. I was very vulnerable and did not object as he put a good plan into action. He decided everything and I just followed. For me, it was good. We reopened in Sweden and I moved to London to continue the business I already had there. After a few years, our partnership came to a bitter end and we sold the business.

This gave me some money, and I went on doing what I always thought was cool—living on a beach in a beautiful sunny place. I moved to Barbados and started a Jet Ski rental business. I soon realized that life on the beach was the same as in an office and had similar problems.

So when the person with whom I was testing a business idea in France six months ago called and informed me that our client wanted to roll out and wondered from where I intended to work. I realized I needed to go back and pursue the business that I was good at. Luckily, I spoke French.

I spent ten years in France, and we built up a big company. We designed our own offices—a 750 square meter wooden building with central heating and rainwater harvesting system. A wooden, ecological building led to a lot of media coverage.

After being away from Sweden for fifteen years, I wanted to move back. I missed my children and grandchildren. So, I sold my part of the company.

It might seem like that everything has been easy, but that's not the case.

Almost everything I have done received criticism. I often ask myself where I got the courage to go after what I want, bearing in mind the people who tried to stop me. I think I am naive. I express what I want. I do not see the obstacles, or if I see them, I do not let them decide my course.

For instance, when I moved from the outskirts of Stockholm into the city, I wanted to live on the street where my mother grew up because I have nice memories of my childhood with my mother and grandparents. I wanted an apartment there with a terrace. This street has old buildings with no terraces. People told me that I would never get one. I found my apartment with an eighteen-square-meter terrace.

Another time when I lived in London, I wanted to live by the sea. I had lived by the sea for a short period in Barbados, but I wanted the same in Europe. I just did not know where and when.

I met with my French business partner at a convention in Paris. He saw my specific skills and asked me to set up a company with him. He lived in northern France by the sea. So my flat happened to be just a few meters from the sea.

While moving back to Sweden, I decided to live in the village where I spent my summer holidays ever since I was a one-year-old girl and I have many pleasant memories. I wanted a land by the lake. But there was no land to buy by the lake. It did not exist. I was determined and kept expressing my desire, and after a while, the real estate agency called and showed me a land—by the lake.

I designed my own house with an indoor swimming pool that has always been my dream. I received several negative comments about how difficult it is to have an indoor swimming pool. Ironically, from people who never had one.

Today, I live in a nice house by the lake, and I own my dream car—a Mercedes convertible. I have a dog—a lovely five-year-old Rhodesian Ridgeback—that I always wanted, gifted by my son.

I am grateful for what I have. I have it because I have been brave. I have always expressed what I wanted. I have not let the negativity affect me nor listened to the naysayers.

I have gone through difficult times, filed for bankruptcy, two divorces, raised two children on my own, and an ex-husband who committed suicide. These challenges never stopped me.

I now look forward to my new career in the network marketing business, which is a new experience for me. When I moved back to Sweden, I wanted to learn something new, meet with new people, and undergo new experiences.

A friend introduced me to the possibility of earning money in a new and exciting way by leading, cooperating, and helping other people to succeed together with me. I have just started this exciting journey, and I am determined to succeed.

Here is my advice:
- believe in yourself
- do what you want to do
- follow your gut feeling
- don't seek all answers in the beginning; eventually, they will show up
- don't be afraid of failing
- be brave and take opportunities as they come
- don't pay heed to negative people
- don't believe everything you see on social media
- base your decisions on your own experiences
- don't harbor preconceived opinions; look for facts
- be kind, helpful, honest, ethical, loyal, trustworthy, and loving
- be persistent—it takes time to build something stable

It is not always easy, but I keep working on my life every day.

BIOGRAPHY

Iréne Wrigstedt is determined, stubborn, and does not mind the obstacles, which has led to achieving her goals. Iréne wants to encourage others to dream big—express your goals, believe in yourself, be brave, and you will get what you want. She writes about life experiences and the consequences of her decisions and actions. Iréne has been a self-employed entrepreneur for thirty years. She introduced new market ideas, built companies in six European countries, had thousands of employees, and helped hundreds of clients to make money. She has made mistakes and failed. But this didn't stop her, and she continued working on new projects. Iréne lives in Sweden. She has two children and grandchildren who are precious to her. She loves her long daily walks with Khathu, her lovely Rhodesian Ridgeback. She plans for a new career in network marketing, along with playing more golf and dancing more salsa.

CHAPTER 16

REACH

By Joan Kenyon-Woods

From my humble beginnings in Jamaica, West Indies, I understood the concept of hard work. As a five-year-old child, I remember caring for my aunt's and uncle's children. I had chores, too, like fetching water from a huge well. There was always the possibility of falling in and drowning, but that thought never crossed my mind because I was given the task of fetching the water for cooking or washing, and I had to complete it. Young, old, everyone became adept at gathering water from that well. Being expected to contribute along with everybody else taught me the value of family, community, hard work, and what it takes to succeed.

We left the idyllic countryside of Jamaica and immigrated to England when I was about six years old. I was enrolled in school, which was not such an ordeal, as I was already accustomed to the British education system that the Jamaican schools had adopted. However, I was not prepared for the class that I was placed in—I was in the third class from the bottom. It was a system that tracked students into homogeneous groups. The school was called Copland, and the classes corresponded to the name, with C being the highest and D the lowest. I was placed in A, the third from the bottom. As I languished in that class, I was acutely aware that I did not belong there. I felt smothered and lost. My constant thought was that there had to be more than this. I knew I had more in me, and I wanted more. I suppose I manifested what happened next.

I loved sports and excelled in track and field. I always gave my all and was very competitive. I could run faster than some of the boys in my school. The British love sports, and track and field is popular. I ran track, and I also played a game called Netball that is the female equivalent to basketball, except you can't move once you have the ball, but only pivot. The objective is the same—you score by getting the ball through a netted hoop. I loved sports because it created a level playing field for me. What I seemed to lack in the classroom, I made up for on the sports field.

For my parents, being in class 7-A, the third from the bottom, meant failure. Like Sisyphus, it seemed that I was rolling the stone of my high school life up a hill destined to have it roll back down into this dead-end of a class. However, belief and a great attitude can change everything.

It was sports day; my time to shine! It rained—that's England, but I was determined that it was not going to rain on my parade. I had a positive attitude and upbeat spirit, though I do not know where it came from. Although, I have to say, my parents were happy people. My dad always made jokes and wanted people to laugh, and my mom loved to laugh.

That sport's day, the gym teacher left the playing field and casually said, "Joan, you are in charge," as teachers did in those days. Nowadays, you dare not leave a class or students to their own devices. It is a recipe for disaster. But, that day, something switched on inside me, like a light bulb coming on. I rallied my team, and we took on the biggest and the toughest.

Although, the tough kids were not going to succumb to the authority of another kid. So, it was a scrappy game. My nemesis, who shall remain nameless, dropped the ball, which I promptly kicked—only to have it hit her right in the face. To say that I thought I was dead meat is an understatement. I had one recourse, she had to catch me first. I didn't have time to think, I just started to run. I knew neither she nor anyone else could catch me. And she didn't. That's the first and last time I ran from a fight. I don't mean a fistfight, hell, I'll run from those every day, but I decided at that moment that I would face life's challenges head-on.

Not long after that, the same teacher—my savior, my mentor, my hero—moved me from class 7-A to 7-C. The challenges of this elevation in life were many. Now, I was required to work twice as hard to get the desired result. The greatest reward, however, was the caliber of people with whom I was now rubbing shoulders. It was as though my whole life had turned around in an instant. I do not remember the people in class 7-A, but, to this day, I have lasting friendships with many of the people in 7-C. What did I learn? I learned about discipline, persistence, and perseverance. My mind expanded and never contracted again. I developed an avid interest in personal development, consuming as many books as I could on the subject; I realized that the world is wide, and I wanted to know how wide. It was not as though everything changed overnight; it took some time to adjust. But adjust I did. I wasn't good at math, so I was in the class below, 7-O. It mattered not. When I was with the movers and shakers, I leveled up. I elevated my thinking, my expectations, and my efforts. The conversations we had, the ideas we discussed, It was as if I was part of a mastermind group. We still have those earth-shattering discussions about politics, relationships, religion, and everything else.

These were brilliant people who I did not know existed until my gym teacher changed my circumstances, for which I will be eternally grateful. I often think of Mrs. Okupa and have tried, in vain, to find her. I remember those little, short, all-white outfits she used to wear, like a tennis player. They were crisp and clean, with precise and well-formed pleats. That was her uniform, and she wore it every day. It seemed as though she must have had hundreds of the same outfit stashed in her wardrobe. I don't remember what part of the continent she originally hailed from, but her African accent was rich and warm, and she exuded a confidence that commanded respect. She was loved by all, but perhaps most of all, by me, because she was responsible for transforming my life and putting me on the path I was to follow. It was Mrs. Okupa who gave me the belief in myself that I have relied on all my life.

Then one night, when I was writing this book and talking to a friend, my phone began ringing. It kept on incessantly, but I didn't answer it—I had that sinking feeling, the one you get when the phone rings late at night—it must be bad news. However, when my aunt, who lives in Canada, and my cousin, who is usually asleep by 8 pm, kept on ringing, I knew it must be something serious. When I finally plucked up the courage to answer, it was to learn the news I had been dreading; my beautiful, lively, generous, intelligent, hard-working mother was dead. Alzheimer's is a cruel disease, and she had been suffering from it for a while. The whole family was forced to watch, appalled, as this wonderful woman literally deteriorated in front of us. Of course, despite knowing her death was likely to happen at any time, I was devastated by the news, and the loss was still very painful.

However, as time passed, my initial sense of panic gave way to the remembrance of all the wonderful lessons she had instilled in me. I remembered, when I languished in that dead-end class, my mom always encouraged me to reach a little further, do a bit more, and aim to be more than I was. When I think of my mother, I think only of her successes. I knew there had been failures in her life because my parents were divorced, something which my mother never really got over. However, I also watched her turn that failure, as she saw it, into a resounding success. She was, by all accounts, a workaholic. That was why my mom ended up owning four houses within ten years of us moving to the USA. It was through lessons such as these that she taught me to reach for what I wanted.

When we immigrated to the United States of America, the move could have been far more traumatic than it was, if not for my mom. We had to leave everything behind in London: my dad, my friends, my way of life, even a boyfriend. But my mom made the transition easier, taking us to live with my aunt, who had us all working within two months after our arrival. We arrived in September, and by November, I was working in Macy's, New York. That was a culture shock. We were all living my mom's dream. I would hear her talk on the telephone to my aunt back in London, who told my mom that the USA would give her everything she wanted—and she was right.

The job at Macy's famous department store was my first ever—and turned out to be quite an important one for me. To get there, I had to travel into Manhattan every day on a crowded, smelly, hot train without having a clue how to navigate my way through an equally crowded Manhattan. I worked in the lingerie department, and enjoyed selling the most expensive nightgowns, those I couldn't afford, to the many husbands and businessmen who asked my opinion.

It was difficult getting up in the morning when it was bitingly cold, but I did it because I knew that my salary helped support our household, and none of us could live for free. I was grateful for the strangers who sometimes gave me their seat on the crowded train. It didn't happen often, but when it did, I was genuinely grateful.

In retail, the adage is that the customer is always right. I don't know if the customer is always right, but I do know that you should always be polite towards them at all times. Customers might often seem annoying, irritating, demanding, and sometimes ridiculous, but, in serving them, we must make a conscious decision to respond positively and be as helpful as possible. Although there are times when we would like to give way to our emotions in such circumstances, it's better to remember that the moment we lose control, we have lost control of the situation.

In retail, one learns to practice mindfulness. The ability to control one's response to external events becomes almost a part of one's DNA and can be the key to success or failure. One has to reach inside oneself and regulate one's emotions, and have some compassion and empathy for oneself. By developing this ability, one gradually learns to decrease the reactivity and mood fluctuations that can occur when one lets events trigger and control one's responses.

One practice that has governed my life is remembering to be grateful. Be grateful for the little and the big things. I remember an old proverb, I don't recall the exact words, but the message was: "Gratefulness can transform the common days into thanksgiving; it has the power to transform life's routine tasks into joyful ones and turn opportunities into blessings." As the author Robert Brault advises, enjoy the small things today because, one day, you may look back and realize they were the big things.

Human beings have an infinite capacity for greatness, so what is it that causes us to fear reaching out for that greatness within us? Is it simply that we fear rejection, the knock-backs, or the ridicule of others? As Franklin D. Roosevelt said, "There is nothing to fear but fear itself." But how does one overcome fear? I think it is simply by taking action to do the thing that one fears. Yes, that is how one overcomes the fear. Let me share an example with you from my own life: I was on a trip to some exotic part of the world and was offered a ziplining activity that consisted of three levels of ziplining. I was eager to have a go, but there was a young lady who had accepted the challenge but, at the eleventh hour, became fearful. I was incredulous, as she was already at the top of the fifty or more steps that we had to climb. There were people of all ages there. I was one of the oldest, and there was a young boy of about ten among the group, too.

I began to talk to this young lady and discovered it was her birthday, which is why I assumed she had taken the trip. Now, she found herself fifty feet in the air, ready to zipline across the jungle. She cringed and held back until the last moment. I shared with the group that it was her birthday, and we began to sing Happy Birthday to her. This completely changed her mood— there is something magical about that song. She smiled and laughed while I gently coaxed her into the harness, and she was made to feel secure. I assured her that I would not leave her and even held her hand. Then, before we could exchange another word, she was whisked off over the jungle. I met her at the next level; there were no more jitters or fear in her eyes. Everyone who had arrived before us cheered and clapped for her and, yes, we sang the birthday song again.

BIOGRAPHY

Joan Kenyon-Woods is a motivational speaker, entrepreneur, and world traveler who exudes positivity. Joan has utilized her professions—as a teacher for sixteen years and a social worker for more than a decade—to gain insights into social and emotional responses to life's varied hills and valleys. Joan was a member of the National Speakers Association, where she rubbed shoulders with the likes of Zig Ziglar and other well-known motivational speakers. A wife for over thirty years and a mother of three, Joan is not only a successful businesswoman but also a loyal friend.

Contact Information
Facebook: http://facebook.com/JoanKenyonWoods
Instagram: https://www.instagram.com/JKWoods_love/

CHAPTER 17

Unleash Your Power Within

By Josef Buchmayr

Austrians are different. At least that's what people tell me when talking about business. I don't believe it. In my experience, people all over the world have the same dreams, desires, and aspirations with one thing in common—time. Everyone has the same 24 hours a day. It is how we treat these hours that makes all the difference.

Let me tell you how everything started for me. As a young boy, I had big plans. If someone asked me then, what I would like to become as an adult, my answer was straight forward: "I will become a millionaire!" My perspective has changed a lot since then. Today, it's no longer just about creating wealth for myself and my family, but to help others in living life on their terms. To be able to do so, for most individuals, includes having the benefit of time and money.

Most people either have one or none of the two. So how do you get both time and money at the same time? You need to own a business. I'm not talking about being self-employed with no employees—doing all the work on your own. I'm talking about being a business owner and having a system that works even if you're not present.

The moment you find a system that works for you and make the decision to pursue that path, your subconsciousness or inner programming needs to be fixed to unleash your power, live up to your maximum potential, and fulfill your wildest dreams.

You have to recognize that you're worth it, and it is okay to live a life of abundance and have more than those around you who weren't willing to persist and pay the price. You don't have to feel bad or have a guilty conscience if you get the success you always dreamed of.

Let me tell you my story. Even though I had big dreams as a kid, I buried them. Only because a person very close to me told me that it was impossible to become a millionaire if you didn't have your first million dollars in your bank account by the early age of eighteen. As a young boy, this seemed out of reach to me, and that's why I listened to my parents and chose the academic path to prepare myself for a good job.

When I was nineteen, I had my first encounter with entrepreneurship. My father introduced me to two businessmen who showed me a venture they were working on. That day, I envisioned a limitless future and got so excited that I couldn't sleep. I began working, but facing the first few challenges, my enthusiasm wavered, and I quit.

I don't recall exactly, but about one or two years later, a good friend of mine called me and pulled me right back in. We worked on a business venture together and had some initial success. Not long after that, things got tough, and my friend decided to no longer partner with me. This was during our student days. This time I didn't quit, but I stopped putting in the effort necessary to advance one's business and got back to student life. Indeed, it was not much different from quitting.

After receiving a master's degree in Business, I decided not to work for someone else, but to live the entrepreneurial life again. The timing wasn't perfect, as the industry I picked was telecommunication with a focus on landline telephony. I worked hard to establish a business, but the product was going out of the market rather than flourishing. Considering the wrong timing, it didn't take long until I ran out of money and had to get a corporate job. I felt devastated and depressed. I lost faith in myself and felt worthless.

Looking back at my career and the several setbacks I have faced so far, I have realized one thing. The greatest enemy of your success is you. It's your subconsciousness sabotaging your efforts. It's your subconsciously thinking that you cannot be more or have more. It's you subconsciously telling yourself that money is hard to be made and that success is something for people better than you. It's your subconsciously believing that you don't deserve the success you're striving for. This is what held me back for so long. When you finally let go is the time when everything will suddenly come together and move in the right direction.

God has a sense of humor. A couple of years later, after I had established myself at a big international company and started earning a decent income, entrepreneurship showed up at my doorstep again. I met with an old friend of mine whom I hadn't seen in a very long time. He mentioned a business idea that he was now pursuing. This is when I pulled myself up and started a new venture. Only this time while working in a demanding job and having a young family to take care of. I believe God wants to test you and see your true character before he allows the universe to be on your side.

But what should you do if you find yourself in a situation with very little time for a new business? I will give you five steps you can follow.

Step 1: Have a Vision

If you have a good and well-paid job, your family and friends won't understand. Most people, including the ones closest to you, don't believe that an affluent life is possible for them, so why should it be for you (even if you have their best interest in mind). In the beginning, they won't support you, or even worse, they will be against you working on your business. To endure and become successful under such circumstances, you must have thick skin. And for that, you need a *vision*. You have to see the future before everyone else does. You must envision a bright future and imagine all the good that will come with it, for you, your family, friends, and all those you influenced.

Step 2: Make Time

As I mentioned earlier, everyone has 24 hours each day, not a minute more or less. So if time is a constant, how do you make time? The truth is, with the decision to do something new, something old has to go. If you're still spending time watching TV, playing video games, or consuming gossip and negative news, these are things you can easily reduce or get rid of completely. Besides that, you can evaluate the work that needs to be done on a regular basis, like household chores or gardening. You could hire someone to do these chores for you.

If, after that, you still need more time, these are the sacrifices that come along with your decision. Family time and hanging out with friends may have to go as well. You must not neglect your friends (your real ones) and especially, your significant other and children. You have to reach an understanding that is okay for everybody. I, for example, made an agreement with my family that Sundays are family days, which means no work from dawn to dusk on Sundays—*family only*. So if occasionally I have to make a conference call on a Sunday, I have to ask permission first. And if a family trip is planned on that day, the call has to be rescheduled.

One tip here: Open up your mind. Instead of telling yourself that it isn't possible to get any more time, ask yourself, how is it possible? This way, your brain will go to work for you and come up with some ideas. You might be able to squeeze in a call on your commute, or you can meet with somebody during your lunch break.

Step 3: Find the Right Things to Do

Now, as you've found a couple of hours every week that you can devote to your business, you shouldn't waste this time. You must work efficiently on getting things done in these hours as humanly possible. This means no distractions and laser-focused work. But just working efficiently doesn't get you your desired results. You also have to pay attention to the tasks you are performing during these hours. In every field of business, there are a handful of tasks you have to do to achieve results; these are often called income-producing activities. Find out the venture-specific tasks and focus solely on one of these during the precious time you have for your business.

Step 4: Create New Habits

Successful people have one common characteristic: they do the things that average people don't. They don't necessarily like every task they do, but if it is necessary for them to achieve their goals, the temporary pain is well worth the payoff that they've visualized in their minds and are soon to experience in reality.

Whether you like the task or not, to do something you're not used to requires willpower, and this will bring down your overall energy level. If you don't like a specific task, it will require even more willpower and drain your energy faster. There is good news. You can relax here. Push yourself to do the activity for ninety consecutive days, every day without fail, and you will create a new habit. After this period, it will even feel awkward if you don't do the activity even for a single day. But don't try to establish too many new habits at once. Pick one habit at a time; otherwise, it might get too overwhelming, and you might give up before the activity turns into a habit.

Step 5: Persist

Work hard and don't look left or right for at least one year. Have faith that despite all possible difficulties, hurdles, and setbacks along your way, you are on the right path, and everything will turn out fine in the end. Keep on going until the success is yours.

Let me conclude with this: we're all on a journey, and we always will be. And because we are more powerful than we believe, it is our duty to always get better and to march forward and make a difference in the world. So if you ever feel overwhelmed, read this fantastic quote and know—*you're worth it!*

"Our greatest fear is not that we are inadequate. Our deepest fear is that we are powerful beyond measure. It is our light, not our darkness that most frightens us. We ask ourselves, who am I to be brilliant, gorgeous, talented, fabulous? Actually, who are you not to be? Your playing small does not serve the world. We were born to manifest the glory of God that is within us. And as we let our own light shine, we unconsciously give other people permission to do the same."

by Marianne Williamson

BIOGRAPHY

Josef Buchmayr holds a master's degree in Business from IMC Krems, a renowned University of Applied Sciences in Austria. He spent time abroad in Poland and the Dominican Republic. He now works for a big international company headquartered in Vienna, being responsible for risk management in Canada and the USA. Besides that, he is building a business in the travel industry and enjoys helping people to live a life full of fun, freedom, and fulfillment. He loves to spend time with his five-year-old daughter and his one-year-old son.

Contact Information
Facebook: https://www.facebook.com/TripleF.sunshine/
Email: buchmayr.marketing@gmail.com

CHAPTER 18

THE POWER OF FAITH
By Leslie Freeman-Wright

Scotland's national poet Robert "Rabbi" Burns famously wrote a poem—*To a Mouse*, in which he wrote the immortal words (translated from the original Scottish): "The best laid plans of mice and men often go awry." These lines are often quoted even today, perhaps because they contain an eternal truth—even our most meticulously laid plans and schemes are often disrupted—by unforeseen things that happen in our lives. Just when you think you've thought of every detail, drawn up a brilliant, fail-safe plan, and set reminders for yourself, somehow that "thing" you'd set your mind on ends up falling through the cracks. When our carefully laid plans come crashing down around our ears, we often feel that all we want to do is run away, maybe to a quiet beach, where we can forget everything and just listen to the waves crashing on the beach while we drift off into *Neverland*.

That feeling is something I completely relate to. In the early part of 2018, I felt everything I'd worked for, along with my marriage, my faith, and my family were slipping away from me. I had no idea what to do and nowhere to escape to; I was lost. Not knowing or understanding how I was doing it, I woke up every day and went through the motions. At the time, I felt as if I was in some sort of coma, standing back and watching my life crash and burn like a downed aircraft. I felt powerless to act in the face of unforeseen life events.

It began when my father passed in July 2018. Within hours of his passing, I landed a whopping six-day period of hospitalization from a disease I'd never even heard of. As a result, I couldn't attend my own father's funeral. As you can imagine, I was devastated. Life as I knew it was already a whirlwind, but suddenly not having Dad around felt as though everything inside me was dead. Those of you who have lost a loved one will understand this feeling. At the same time, I realized that my marriage was failing, and felt there was no one to turn to share the burden. I was in a complete daze. Just like most people going through such events, I began to realize just how precious and short life is.

Prior to this, I had dreams and aspirations of becoming "someone." Back then, I didn't understand that I was already "someone." I had to dig deep inside myself to find the strength, the motivation, and faith in God's will to make it through the days. Fortunately, my husband and I realized we wanted to save our marriage and make it stronger. We discussed it and decided that we'd create an entirely new marriage. We both understood that no one's perfect, and that we needed to regroup and go back to the time when we initially said "I do" to each other. Part of the problem was that we both have such busy schedules, but we knew we had to work out a way forward and make a decision. That decision was to *work at working* on our marriage. At that point, it was like God tapped on my shoulder and gave me the tools that we needed to rebuild our life together. It wasn't easy, but it was worth it.

You see, I realized that I was placing my faith in all the wrong areas at first. By definition, faith is having complete confidence and trust in someone or something. When I looked closer at my life, I saw that all along, I'd been hoping for different outcomes in most areas of my life, but somehow expected them to occur miraculously, without any effort on my part. I wasn't putting in the work and effort to make the changes necessary to bring about those wished-for outcomes. Instead, I questioned why things, both good and bad, were happening, but never took the time to take a step back and try to understand the reasons for them. However, there came a moment when I finally slowed down and took a breath; that pause gave me the chance to see that everything happening was more significant than me and that I alone wasn't going to be able to change it. For once, I had to make myself believe that I wasn't a terrible person and that my marriage wasn't failing. That horrible feeling of being in freefall, that everything you think you have a hold on is disintegrating, might actually relate to things falling into place. By looking at things from a different perspective, I began to understand that the trials and tribulations I was going through might actually be in my favor. Hard to believe, isn't it? This is when the phrase "with God all things are possible" came to mind, which helped me to understand that the faith factor I was missing was ME. I'd given up on me a long time ago, and it took my world to fall apart for me to look up and see that I believed in God and his faith in me, but I needed to rebuild myself and my faith within if I was going to move forward.

When I sit back and look at every major obstacle that was thrown in my direction during that dark time, I now see it as a period of preparation: Preparation for what was to come and finding the way to getting where I wanted to go. I believe that all things happen for a reason, and light shines on every shadow; I'm here to testify that believing in myself and what I could do with passion and positivity set forth in my life, and a change of mindset is powerful.

With this new insight, I fought for my marriage; I went to individual and joint counseling with my husband. Together we created a prayer wall, read a lot of relationship guide books, and went to a marriage retreat. There we learned valuable life lessons that not only helped with our relationship but also to become the positive role models our two beautiful girls needed. Talk about relationship goals! I'm proud to say that through our fight, determination, efforts, and hard work, we are now one hell of a power couple, and I'm so glad our kids have strong role models to look up to.

Life wasn't easy, and trying to rebuild it from the ground up was a definite struggle at times. There were tears, yelling, breakdowns and the "give up" attitude. I fell victim to that situation where you wake up every day wondering what else was going to happen. Many times, I just wanted to close the chapter and walk away. I wondered how things had got here, where did we go wrong? Yet, at the same time, I had to dig deep, stand still, take a deep breath, and start the day anew, and keep pushing on to find the belief that I'd buried deep inside. One of my sanctuaries was going to my fathers' gravesite. There were days when I'd just park my car alongside the cemetery and cry while playing some of Dad's favorite music. Though he wasn't physically here on earth, he was still there for me. Anytime something new happened in life, good or bad, I'd make a point of going to hang out with him, make sure he was okay, and share my stories with him. Of course, even today, it still breaks my heart that I wasn't able to attend his funeral because I'm sick with a disability that's invisible to the naked eye.

At times I felt angry, questioning God as to why these things were happening to me, asking where did this disease I'm afflicted with come from, and why now. I'm not going to lie; that was by far the most difficult part of my grieving, not being there for my dad. However, as with all things, I know he wouldn't stand for me, NOT getting the proper treatment that I needed. Yes, all of this was going at once, and good days were hard to come by.

But, through these struggles, I somehow managed to find out what a strong person I am. I might have been broken, or so I thought, but I never let that keep me down. Somehow, through the pain and the sadness, I mustered the energy to pull through. I believe staying in tune with myself and the new version of who I wanted to be made all the difference. I refused to let the bad win. I'm a fighter, and I would prevail. Aside from my past and things I couldn't change, I could still learn from what was happening. I learned from my mistakes, so I knew what NOT to do moving forward. I had to fix my mindset and remain as positive as I absolutely could. My energy spilled over into my relationship, and my husband and I began to implement changes, putting into action everything we'd learned over the years, so that our love began to thrive once again. I could have walked away numerous times, but that is the easy way out. Easy for the relationship, but hard for the kids affected by their parents splitting. But again, learning to fight for what you believe in, and letting faith intervene was the blessing that I needed.

What would you do in time of need? Would you take the easy route and walk away, or would you stay and work for a better lifestyle? Would making changes to your daily habits benefit your overall health, both physically and mentally? If they would, how bad would changing your mindset, life, and routines be if you knew the outcome would be beneficial? What if you had no idea what was on the other side? Would you make the leap of faith and believe in yourself to see what the outcome would be?

Everyone has a story, a struggle, a romance, a heartbreak, but not everyone is strong enough to pull through them. Life has a sense of humor where you first get the test, then the lesson comes later. It's kind of crazy how that works. What you don't realize is that everything seems important—until you get sick, whether that's physically or mentally. For it's at such times that you realize what's really important to you. Believing in your self helps you to find a way to keep going and rebuild your life, to leave behind whatever has you stuck.

Most importantly, you'll learn to take care of your health and start to realize that whatever you were fighting against might be the very thing you need to work on to make it right. Sometimes, you just need a moment to exhale, take a good look in the mirror, and realize what you have and what you're working with. If you alter just one thing every day, you'll see what God can do. Believe in yourself, because the power of faith is so strong. If you just believe in yourself, as I learned to do, and alter one thing at a time, your mindset will change, too. My experience proves that all great things happen to those who work hard for them and focus on believing that faith lies within themselves.

To date, my husband is my rock, my confidant, and my best friend. We aren't perfect, but the Lord knows we were meant to be joined as one. Together, we're working towards our future goals and building a proud legacy that our girls can one day emulate. It's all about the right mindset, belief, and faith. You don't have to believe in God. Just remember, if you don't have faith in yourself as a person, then no one else will. There are times in life when you just know that you have to do what your heart tells you, when you decide to take a chance on fulfilling your longheld dreams, even if no one else can see that. Going through these tough times taught me that the power to take control of my own destiny was in my hands— I had a vision, and I didn't want to lose it. Time after time, I faced seemingly insurmountable odds and stumbled, but I reminded myself that I'm in charge of my own destiny, I am a mother to two young girls who need a strong role model, a wise leader, and supporter. If I wasn't healthy both mentally and physically within my relationship, then I was a failure, something that I never wanted to be. I knew that if I didn't make changes and uplift myself to find my own happiness, then I couldn't expect those around me to look at me as a source of strength and positive motivation. We all fall, but it's up to you and the faith you have inside you, to choose whether or not you get up, and how you do it.

BIOGRAPHY

Army Brat turned Army Wife, Leslie Freeman-Wright, tells the story of her life and her journey to becoming an important role model for her family, as well as in her community. Working for the airlines, Leslie travels to many places for training events, learning, and teaching others about the industry and how to become successful. In addition, Leslie has been an entrepreneur for over ten years, stepping out into the world away from her day job, and experiencing other things that life has to offer. She has always wanted to write her own books, and finally found her niche through training others to find their full potential in life. By doing so, she has found happiness by helping others succeed in everything they want and love to do. Leslie has also been an advocate speaker at several training seminars and continues to lead everyone she meets find happiness and success.

Contact Information
Facebook: https://www.facebook.com/leslie.freemanwright
Instagram: https://www.instagram.com/poison3f98/

CHAPTER 19

How Moms Can Change Lives And Improve The World

By Maria Helena Paulo

I am about to share with you how the worst pain I have ever experienced transformed me into a new and better version of me.

I lost my dad in January 2020 after having lost three of my younger brothers over the previous eight years. It is so painful to lose family and, at the time, I lost focus on everything. I stopped going out with my best friends, which I love to do. I was mentally blocked. I didn't know what to do. I wasn't sleeping enough, not eating, not talking. I had no direction, and I was frustrated.

Being the firstborn daughter, I had to care about my mom and the rest of my family, but I had no energy. Then, one day, I went to Mom's house, and we talked; I found that she was so strong and the only one who didn't need my support. She is so much stronger than me, and I dedicate this chapter to my mom, Luisa.

After around three months of losing Dad, I decided to turn my life around, to renew my life. I started rereading books on personal development and happened to read an article by Sashin Govender, a young millionaire student from Durban,

South Africa. He recommended that I read a book by one of his best coaches, Matt Morris.

I am retired from the banking industry, and since August 2018, I have been doing part-time network marketing to help out my mom financially. In the three months since losing my dad, I had lost almost two-thirds of my earnings and a big part of my team because I was not there to work with them. I had lost all my motivation to do anything, and my level of leadership and authority had deteriorated.

So, by reading various articles, I finally found a book subtitled something like: "I was able to finally learn the secrets of stepping up to a powerful position of authority, and my income skyrocketed!" It was by Matt Morris, who promised to teach me "exactly what those secrets are that almost nobody knows about. One: It's in your hands; this book can take any entrepreneur from beginner to celebrity status practically overnight."

From that moment, it was as though something clicked inside me, in my heart and in my mind. I thought I was dreaming, that it couldn't be real. I even wondered if I was having a vision! So, I decided to purchase the book that promised seven secrets to a seven-figure income, but something went wrong during the buying process.

Then, on 7th May, 2020, I received a message from Matt Morris that said: "Maria Paulo, what happened? I noticed that you tried to purchase my book a few hours ago, but, unfortunately, it looks like your order didn't go through. I wanted to check-in and make sure all your questions are answered."

I thought, my God! How can this be real—a celebrity writing to me!? So, I decided to finish the payment process and purchased the e-book. I downloaded, I read, and I watched the videos and the MLM Masterclass audios.

On 8th May, I saw that there was an invitation in my online mailbox to become a co-author to improve the chances of publishing success. I duly filled in the application form but was not expecting a positive answer, as I was depressed and had no experience in writing.

I had never written a book, but I remember having a strong vision of doing something big for my mom, because if she'd' decided to get pregnant during difficult times in 1956, then I was not going to let her down by failing to answer the call of a millionaire, who was willing to teach me how to win big, just by making a decision. That's another reason for being grateful to my mom, Luisa. Dreams really can come true: I remember being alone in my room when I opened my mail on the 14th of May and listening to Matt in his video saying, "Hi, Maria, Congrats for being selected as part of the team of co-authors."

Again, I did not believe that he was actually talking to me!

Moms produce many successful people, people who are leading the world, millionaires who are positively impacting the lives of other moms and their children. These people are scientists, entrepreneurs, constructors, and healthcare workers. And they do this by merely deciding to be moms, by transforming that decision into action. They've found the "how-to" of it, they are committed to their decision, and they care for their children. They are motivated, they believe in themselves, and they use their power positively.

Finally, I say that the world should thank all moms because, by their decisions and commitment, they can change peoples' lives for the better, so that there are smiling faces all over.

Now, let me share with you how far moms can continue doing big things.

Moms, you know how your children are always watching what you do as an example. As you want the best life for them, you should make sure that they see you doing BIG things.

They don't do what you say, but what you do!

Moms sometimes have to struggle to empower their children, but I say, "Woman, go and have freedom of lifestyle; you're powerful, you do have a better way, so go and show the world!"

Moms should encourage other women to do network marketing, as it's a perfect fit for stay-at-home moms. So, if you're a mom who can imagine and believe that it's possible to become a millionaire, you can bestow freedom and help people.

It's a fact that many women are in business but are not making enough money. One way to start making money is through network marketing, where you make a small initial investment on joining.

In the network marketing industry, there's no age limit, no gender, race, or religious bar, and you don't need a formal education, certificates, or degrees. What you do need is to have a vision, to set up your goals, and surround yourself with positive and successful people.

Moms, what is required is to focus on following the system, on doing what your mentors do, constantly and persistently.

There's only one question you need to ask your customers: Do you want to have extra money? If they say yes, then you close the deal with them, and you earn your commissions by spreading your company worldwide.

Moms, you also need to commit to daily habits, like more time, more freedom, and a lot of money.

Moms, you can and should raise your children, and by doing it, you can transform people, helping them toward better health, wealth, and relationships.

When you feel tired or depressed, you can turn that around and fuel yourself with personal development through seminars, webinars, and reading books.

What I've learned is that we should dream big, think big, and realize that to meet our goals and achieve our dreams, we first have to master our skills. To get the big results you want, you have to put in a lot of hard work.

Another lesson to keep in mind is the old rule of 80/20: only twenty percent of people are seriously working; the other forty percent do nothing; the remaining forty percent do a little. In my experience, it's advisable to foster the discipline of always learning something new to get good results. So, Moms, where do you position yourself within the old 80/20 rule? I have chosen to be part of the twenty percent.

Moms, you should schedule time for exercise because it's good for your health. It may be tough at times, but if you set yourself achievable goals, you will be able to push through, and you'll find it's well worth it. To positively impact the world, you must first improve your self-esteem, be kind to yourself. You don't have to be perfect, but just be the best version of you that you can be. Forgive yourself for your mistakes and protect your mental health. Don't let yourself get lost in a maze of anxiety and fear about things, like the coronavirus, where you don't have all the facts—don't listen to rumor or fake news—get out there and get the facts by listening to the health experts. Then you will be able to forge the best path ahead.

Successful moms start by creating a daily schedule for outdoor time, academic time, home cheers, free time, bedtime, and so on. Moms can create more freedom for their sons and daughters and inspire them through network marketing. Moms have transitioned from working in traditional jobs to building businesses that provide a high-quality lifestyle for their families.

These moms manage their time so that they can be as productive as possible when working from home. They create agreements with kids and spouses that make it much easier for them to grow a business, while still having quality family time. So, again, moms can impact their kids in positive ways, as they build their business by supporting and inspiring them.

I am so grateful for this realization. There are so many moms involved in business, whether working corporate jobs or stay-at-home jobs. Here are some of the solutions to the challenges they face: negotiating agreements with kids, partners, and spouses, traveling together, watching videos together about success stories, or involving your significant ones.

It's critical that your kids see the work that you are doing and the positive results it produces.

Reasons for moms to choose network marketing careers:

1. Moms are looking for flexible hours; you can do network marketing online, via phone, in person, during evenings.

2. Women are usually big supporters of other women and are big on motivating them to succeed, but, in a corporate environment, they often experience fear, criticism, or envy. In contrast, network marketing fosters an atmosphere of love, encouragement, affirmation, and supportive coaching.

3. Spending time around other successful women is exciting and inspiring. After spending time with kids, it's great to have a little bit of time with other moms, but let it be with moms who are both inspired and inspiring and love to have fun while making money.

4. Moms are experts in getting the best value for their money, and network marketing startup costs are generally low. Plus, from your very first sale,

you're already making money.

5. There are no barriers to entry in network marketing, and with moms bringing their experience and know-how from all walks of life, it's the perfect place for them to cultivate faith, improve their level of education and work experience.

6. Moms in network marketing enjoy personal transformation, but also build trust networks, get lots of on the job business training, as well as benefiting from attending workshops and webinars. I guarantee that, within two years in the network marketing industry, you will undergo a personal transformation that will be worth far more than the extra money you gain. The material things you'll gain are not as valuable as the personal growth and sense of fulfillment you'll experience. It's not what you get, but who you become—the best version of yourself.

7. Moms like to lead lives of financial independence where they feel they're making a contribution.

So, you can see that network marketing simply ticks all the boxes for moms looking to improve their own lives and those of their children and families. It's an open, flexible business model that promises rich personal and financial rewards. If you want to learn more, I recommend reading Lauren Kinghorn's book, *7 Rocking Reasons Moms Chose Network Marketing Careers*. All mothers, with their work and personal investment in their children, whether they are in paid employment or not, make a vital contribution to the economy of their country. They represent a huge and important part of a nation's workforce and productivity and are vital to the economy's health.

Therefore, increasing opportunities for working mothers would boost the economy. Coronavirus has impacted the economy, especially for moms working from home, who are stressed about not being able to engage effectively at work. Clearly, this will cost the economy billions of dollars.

With COVID-19, there are pros and cons to stay-at-home parenting.

Pros:
   i. Increase in child's school performance
   ii. The child has less stress and aggression
   iii. Social approval

Cons:
   i. Desire to return to work
   ii. Higher levels of stress
   iii. Social isolation for mothers

I would like to conclude this chapter by encouraging all moms to stay caring and work on themselves, their children, and families for the benefit of the world.

Today, we know a lot more about the great things that great women have done and continue to do; they're businesswomen, family carers, students, professionals, domestic goddesses—and moms!

While we often recognize the significant contributions that women make to society as successful doctors, politicians, and lawyers, as well as in all professional walks of life, their overall contribution to a well-balanced and civilized society through motherhood and caring for others is often overlooked,

To close, I can say all that cleaning work that mostly women end up doing, is essential work because without good cleaning and sanitation everybody gets sick!

There is a light inside each one of us that must be nourished, that needs to be cared for to shine brightly, but sometimes it is overlooked, it has not spoken, it has not loved.

Give it the music it needs to dance; that light is you, it's within you, but sometimes it gets forgotten, yet it's the thing that others most want to see, so let it shine.

If you shine, then you give others permission to do the same. So, always climb up and clean the light itself. Always take care of the light within. Don't hide the light. The world needs you to shine.

BIOGRAPHY

Maria Helena Paulo is 63 years old. She was born in Xai Xai, a small village in Mozambique. She has a Master's degree in economics and has had careers in the civil service for eighteen years and the banking industry for fifteen years. She loves teaching and has spent fifteen years teaching economics and finance part-time at her local university. She retired from the Central bank in 2011 but continues to serve the banking industry as an executive board member. Two years ago, she joined the Worldventures company to travel with family and friends. She loves music, dancing, and helping people. She loves her mom, Luisa, and is married to Candido; they have two daughters—Sonia and Sheila, and are grandparents to two lively boys—Caio and Kizua.

Contact Information
Facebook: https://www.facebook.com/mariahelena.paulo.1232

CHAPTER 20

THE KEY FOR LIMITLESS SUCCESS

By Mike Howren

It was November of 2012 when I received a phone call from an old friend. I can't recall what she said, exactly, but I knew that she was more excited than I'd ever seen her. I do remember that she told me she had something I had to see.

Once she arrived, she pulled up a website on my computer, started playing a video, and I was rolling my eyes within 30 seconds. "Are you pitching me a pyramid scheme right now?" I asked. "I've seen dozens of these companies and presentations, and they don't work. They're a scam and a waste of money. Only the people who join in the beginning make money."

You have probably said the same thing at some point in your life, or maybe you know someone who shared a similar story. I had no exposure to the network marketing industry at the time. Growing up, I was told the key to success was to go to school, get good grades, and find a safe and secure job. So, I went to Kent State University, got my degree in business administration, and entered corporate America. I started working with a marketing company and worked my way up through the ranks over the years. By my mid-twenties, I had become my company's vice-president of sales. There I was with a great job, a great title, and a six-figure income. "Man, I've made it!" I thought, but as the years went by, I took on more responsibilities, which meant more time at the office, more stress, and less time to do the things I wanted to do with the people about whom I cared.

We all know people like that—they make good money, but they have no time because they are always working. We also know people who have a lot of time on their hands, but they're broke. I've always been interested in learning how I could have both time and money. I was at a point in my life where every decision I made came down to two things: could I afford it, and do I have the time to do it? Despite my good job, I was still living paycheck-to-paycheck, was stressed out, and time-broke. I needed a change.

So, back to that day in November. I finished watching the network marketing presentation. Honestly, I only watched it to be polite to my friend, and I had every intention of telling her no. In fact, six months prior, I had actually seen a presentation from the same company and had said no at the time. Well, this time, I was introduced to a very successful guy who became a huge mentor in my life. A mentor—imagine that. I had never even considered seeking out a mentor before. I was too focused on just getting through the week, paying my bills, and waiting for the weekend. That day, he asked me a question that completely altered my path in life: "Mike, how many wealthy employees have you ever met?" he asked.

Now, at the time, I thought that being wealthy just meant having a lot of money, so I responded rather quickly. "Doctors," I said.

He looked at me and said, "Doctors are broke."

I was confused, and I politely asked, "What in the hell are you talking about? Doctors make great money," I said.

"Don't get me wrong," he said, "They might make great money, but they're time-broke. If they don't show up at the hospital, they don't get paid. The person who has true wealth is the person who owns the hospital. He makes money whether he lifts his head off his pillow or not." He then proceeded to teach me two laws of wealth that have changed my life forever:

Wealth Law #1: You need to have multiple streams of income—not multiple jobs but multiple streams. Today's average millionaire has seven.

Wealth Law #2: If you don't find a way to make money while you sleep, you will have to keep working for your money, and you'll end up working until you die.

This was drastically different advice than I'd received growing up. Getting a college education and a good job was all that I'd heard about.

Two things dawned on me that day: one—as long as I was trading time for money, I could never have both; and two—the teachers from whom I'd learned in school didn't have their own financial freedom. How could they teach me about financial freedom if they didn't have it themselves? That day, I made a decision to pay attention to that guy because he had the life I wanted, and that has made all the difference in the world.

I started my network marketing career that day with high hopes and aspirations. I was going to "get rich quick." I learned the presentation and began sharing that vision with my friends and family. Some joined me, but many did not, and my excitement began to fade. The negative tapes started to play over and over again in my head, saying, "You're not cut out for this. You're not the type of person who can have success with this. You have a great job, just focus on that. This is for people who can't find a good job." Months passed, and I was very frustrated and on the verge of giving up. It was then that I had a conversation with my mentor, and he convinced me to go to the company's next major training event. I bought a ticket but reluctantly so. I missed a great friend's wedding, but I knew that I had to figure this out if my life was ever going to change.

That weekend, I saw what the industry was all about, and I fell in love. That weekend, I realized something I saw Grant Cardone post on social media years later:

Everyone should be involved in network marketing.

1. For the personal development
2. For the product
3. For the money

Up until that point, I had the order reversed. I was trying to make a bunch of money, but I wasn't working on myself. I had no clue what personal development was. The last book I'd read was in college. If you don't know, Grant Cardone has built a $1.4 billion empire in real estate, was named the number one marketer to watch by *Forbes Magazine,* and is the founder of the 10X Growth Conference, the world's largest business and entrepreneurial conference. He might have just a little more business credibility than your Uncle Buck, who tells you that network marketing is a scam.

I learned something at that event that changed my entire philosophy in life: your income will always follow your level of personal development; up or down, it will follow. Have you ever heard of someone winning millions in the lottery, only to find themselves broke a few short years later? They did not learn how to become the type of person necessary to earn that kind of money, and it left them almost as quickly as it had arrived. Have you ever seen a millionaire lose all of his money only to become a millionaire again a few short years later? This is due to the self-image these individuals have. A person who is broke views himself as broke, and no matter how much money may fall into his hands, he will spend it until he returns to where his true self-image lies—that of a broke person. On the other hand, a rich person views himself as rich, so no matter what losses he experiences, he will do everything possible to get back to his true self-image—that of a rich person.

We function the same way as a thermostat in our homes. We all have financial thermostats programmed for us, and we aren't even aware of it. Our thermostats are programmed from years of us believing certain things about ourselves. What you believe about yourself is a byproduct of your environment, the people around whom you spend the most time, and the programming of your brain by you and those around you over the course of your life. I wasn't trained on this, not in my entire career in corporate America. Instead, I had to learn the skills necessary for completing my job, knowing that I would be fired if I didn't do my job correctly. My company couldn't care less how I felt about myself—they wanted only one thing: performance.

In the world of personal development in network marketing, my self-image changed. I surrounded myself with people who built me up instead of tearing me down. I read books on personal development and was surprised to find them interesting. These books were drastically different than those I'd read in school.

Network marketing and the personal growth I have experienced has changed every aspect of my life. I've made life-long friends, have created countless memories and had invaluable experiences, have traveled the world, and lived life on my own terms. I've been introduced to a group of peers who have pushed me and made me better. I've had mentorship from some of the most successful people I've ever met. I've become a better leader, performer, communicator, partner, friend, son, brother, servant, and an overall better person. Because of this industry, I've learned how to be a business owner, and I have built a multi-million dollar business outside of my network marketing business. I've expanded my mindset and vision and have created and achieved goals that I previously thought impossible.

So, the next time you get a call from a friend or an old acquaintance who has something he wants to share with you, know this: he is not trying to sell you anything. Instead, he has experienced something that has changed his life, and he thought enough of you to pick up the phone to call you. Answer the call and hear him out. You never know if that one phone call might forever change the direction of your life.

Limitless success is never-ending, and it is available to anyone who wishes to unlock it. The key, however, lies within your mind, and the best way for an average person to access that key is to plug into the system of personal development offered through network marketing.

BIOGRAPHY

Mike Howren is a Northeast Ohio native, traditional business owner, six-figure network marketing professional, and an author. Having grown up in the rust-belt, Mike learned the importance of a strong work ethic. It is this work ethic and discipline that has carried him through the ranks of corporate America, but it wasn't until he started working harder on himself than he did on his career that he started to see a change in his lifestyle. His passions include traveling with his fiancée, Lindsay, and spending time with his family, golfing, and boating. His current mission in life is to teach others and inspire them to live life on their own terms.

Contact Information
Facebook: https://www.facebook.com/mike.howren
Instagram: https://www.instagram.com/mikehowren/
Website: http://www.michaelhowren.com

CHAPTER 21

LIVING THE DREAM

By Richard Denning

Everyone dreams, it's the way God made us. We dream about the future, we dream about how things are going to turn out, the great job we are going to have, the vacations, the big house, great family, and plenty of money. We dream about the things we are going to accomplish, how we are going to change the world and make a difference. All those things can come true if you have the right mindset. If you are willing to do today what others aren't, you will have tomorrow what others can't.

Everyone has their own idea of what it means to "Live the Dream." For some, it means financial freedom or the freedom of more quality time. For others, it's owning their own business, being their own boss, or working at something that they feel passionate about. God has given me a passion for helping other people. That's why I have written this book—to help you to discover what "Living the Dream" means for you. I hope that, after you've read it, a whole new world opens up for you and that you will learn how to dream again.

Before we get into the good stuff, let me tell you a little bit about myself. My name is Richard Denning. My wife is Timberly. We have two children and two grandchildren. I live in northern middle Tennessee, about one hour north of Nashville, Tennessee, in the Fairfield community. I have lived here all my life. I am a fulltime farmer and, over the years, I have raised, cattle, hogs, pheasants, quail, chukar, bird dogs, rabbits, corn, soybeans, wheat, oats, barley, sunflowers, milo, tobacco, and, in 2019, I started growing hemp for CBD oil. I also own a hunting preserve and a ten-acre pay lake. Diversification is key to being successful on the farm. If one or two of the commodities are down, some are steady, and some, hopefully, are up. Farming is a great life, but it takes a lot of hard work—precise planning, guts, nerves, and prayers that Mother Nature will be kind to us throughout the year. Farming is probably the biggest gamble in life. We roll the dice every year. It has been said, "Give a farmer a dollar and he will spend ten." Farmers are probably the most optimistic people in the world.

My entrepreneurship began as a young boy when I started raising rabbits to sell. I would also ride my three-speed bike around the area and pick up discarded coke bottles, then take them to the store and get five cents per bottle, which was pretty good money for a young boy. I bought a lot of Super Bubble, chewing gum, cokes, and peanuts. Some of you know what I mean. I also worked at the high school basketball games for my dad, where I started collecting silver coins. I would trade my money for the coins and one-half dollars. My dad taught high school agriculture for over thirty years, and my mom taught elementary school for more than forty years. As I grew up on the farm, I learned hard work and commitment from both Mom and Dad. But life on the farm was not all work—there was play, too. For instance, Dad built a baseball field where we practiced Little League and Babe Ruth baseball and then played church league softball for several years. We hunted, we fished, and attended church every Sunday. To us, that was "Living the Dream."

I played football, basketball, and baseball all through junior high and high school. I graduated valedictorian. I had big dreams. I had to decide to either go to college or stay on the farm. I decided to stay on the farm. I had lived the dream all through school. Now one chapter of my life was ending, and a new one was beginning. I got married, my wife was going to college, and, for the first time, I had bills to pay, real bills. I wasn't afraid of hard work. I was young, healthy, and full of energy; I was ready to take on the world.

Farming is a rollercoaster ride. Like any profession, it has its ups and downs. I thought that I was "Living the Dream," but there were times when I had to question that assumption. Sure, I was my own boss, owned my own business, and I could take off anytime I wanted. Except that I didn't; I hadn't had a vacation in years. Wow! I realized that my business owned me, and I didn't even know it. I loved what I was doing, but I was getting older, and farming was getting harder. It wasn't as much fun as it used to be. The decisions were getting more complicated—payroll, insurance, and on and on. I had to continue to increase my farming operation to survive. It seemed never ending, and it was becoming more demanding, while profits were getting smaller.

But I was always looking for more income through new ideas. So, when my friend invited me to take a look at an idea that she was excited about, I thought I was too busy. I'd seen those things before. They don't work, and I didn't want to sell products. I almost didn't go along to find out more, but I am so glad that I did because I was pleasantly surprised. I saw a way to save a lot of money and create a really good residual income with the help of a team. I knew about teamwork. I had been on many teams during my lifetime, and I knew that a team could accomplish more than one individual. The residual income really got my attention. I saw a way to cut back on my farming operation and supplement my retirement. Now my dreams were beginning to change. The idea of "Living the Dream" began to take on a whole new meaning.

Warren Buffet says, "If you don't find a way to make money while you sleep, you will work until you die."

Do you want to work until you die? Do you want to do what you are doing now for the rest of your life? What's going to change in the next year, the next five years, ten years, or twenty years? Are you dreaming? Sometimes your dreams change. In my case, my first dream was to be debt-free and have enough money to buy anything I wanted. I accomplished that. But I had to put in sixty, seventy, eighty hours a week to have that lifestyle. Because of that, I missed so much time with my kids growing up. I was time broke. I had created a monster that I couldn't get away from. Farming was taking all my time. I had been taught to work hard, do your best, and you will make it. That's what we are taught in school. That's what society teaches us, the forty, forty, forty plan: Work forty hours a week, for forty years, and retire on forty percent of the income you were making. But that's not enough to live on. Besides, during those forty years, you are helping build someone else's DREAM! Why not start building your own DREAM?

I am writing this during the COVID-19 pandemic. Many people have been laid off or lost their jobs. A second stream of income would have made it much easier for them to deal with the situation. Money is not everything, but it sure does take the rough edges off.

"Dreams don't work unless you do!"

Most of us start with big dreams, then life happens. We go to work, help build someone else's dream, come home, eat, and go to bed. We get stuck in an endless, repeating cycle. Over time, our dreams begin to fade away and get crushed by life. Then there are the others who tell us that we're crazy for having a dream, saying, "You can't do that!" They don't dream, and they don't want you to dream either. NOTE: DON'T LISTEN TO THEM! They don't pay your bills. A wise man once said, "Sometimes the things we regret most in life are the risks that we never took."

What does "Living the Dream" mean? Is it possible? I believe that network marketing is the vehicle to accomplishing your dreams.

When I was first introduced to network marketing, I was working sixty to eighty hours a week. I had no time to spare. But, you know, we make time for what we want to do, and for what is important. I looked at several network marketing opportunities, but none of them seemed to be the right fit. I didn't really want to sell products. So, when my friend asked us to look at something that she was excited about, my wife and I were a little reluctant. But along we went, and we were really surprised at what we found. This time, it seemed to be the right fit and the right timing for us. NOTE: Find something that you love doing, and you won't work another day in your life.

Here's what happened to me: I had "Lived the Dream," but it had faded away until I was no longer dreaming, but I didn't even notice; I was consumed with farming. I was caught up in the rat race of life. Realizing this, I was open to looking at a new opportunity, and what I saw was a way to change my life for the better. This network marketing idea was exciting, and I began to dream again. For the first time in years, I was excited. I saw and talked to regular people who had transformed their lives from a mundane lifestyle to an exciting adventure. I knew that that was what I wanted.

It started with a thought; "If they can do it, I can do it." The wheels in my mind began to spin, and I couldn't sleep for a week. Everything starts in the mind, with belief. If you can see it and you believe it, you can achieve it. The mind is a powerful thing: You become what you think about. I began to think positively. I learned about personal growth, which I had never heard of before. I started reading books like *Think and Grow Rich*, *The Magic of Thinking Big*, and *Rich Dad Poor Dad*. I started watching videos of successful people. Personal growth has helped me in every area of life. I was so excited that I was going to sign up everyone I knew. Surely, they would be as excited as I was? But that wasn't how it happened, and I started to get knocked down. How many times do you keep getting back up? One more time? I love the great outdoors and like to deer hunt. I learned it from sitting patiently in the tree stand in the darkest hour just before dawn. The Bible talks about a time for everything, but it doesn't mention anything about a time to quit. Don't quit! Take it from one of the greatest basketball players of all time: "I missed more than 9000 shots in my career. I've lost almost 300 games. Twenty-six times I've been trusted to take the game-winning shot and missed. I've failed over and over again in my life. And that is why I succeeded."—Michael Jordan.

With the help of others, you can get back up. You can succeed. Don't quit until the miracle happens.

FINAL THOUGHTS

There's never a wrong time to start dreaming. Go for it!

Be open to new ideas. There's one that is right for you!

Nothing happens until someone gets excited. Be Excited! Be all in!

Do you want to change your life and your family's life for the better? Do you want to leave a positive legacy that will last for years after you are gone? If you do, you will have to take some risks. Remember, I mentioned earlier, "Often, the things that we regret most in life are the risks we were never willing to take."

Let me leave you with this thought:

Life is short, so don't let the sun go down on you before you "LIVE YOUR DREAM."

To God, Be the Glory, for with Him, all things are possible.

BIOGRAPHY

Richard Denning was born and raised on a farm in middle Tennessee, the home of country music, and has over forty years of farming experience. He loves to hunt, fish, play sports, and be in the great outdoors. He became a full-time farmer right after high school, helping feed the world. Over the years, he has also created an outstanding hunting business, Meadow Brook Game Farm, where he has met people from all over the world. Creating multiple streams of income has been instrumental in being a successful farmer. Richard has a passion for helping others. In this book, he talks about *Living the Dream* and how his farming experience has helped him successfully launch his network marketing business and allowed him the opportunity to show others how to create an extra stream of residual income.

Contact Information
Facebook: https://www.facebook.com/richard.denning.716

CHAPTER 22

THE ULTIMATE SUCCESS FORMULA

By Romacio Fulcher

In "The Ultimate Success Formula," my goal is to provide you with something simple yet profound that will help you in every area of your life, be it spirituality, finances, relationships, health, or your community. I will give you access to a formula that has never failed to work for me, no matter the situation.

I graduated from high school with a 2.5 GPA. I dropped out of college after 1.5 years, so I can tell you firsthand that I'm not the sharpest tool in the shed. In my 42 years on earth, the one thing I can tell you is that I'm really good at keeping things stupidly simple. The world we live in today is so complicated. There is so much going on that it is easy to feel as if you are drowning, lost, or confused. My aim today is to share a simple formula with you that is sure to work in every area of your life. I am excited to impart this to you, but I do ask one thing in return—when you read this, I want you to apply it instantly and be blessed because of it.

Let us talk about my Ultimate Success Formula. The definition of success, according to the dictionary, is the progressive realization of a worthwhile goal or worthy ideal. This means that the key to being successful is always to be sure that you are making progress. The reason why many of us become trapped and entangled in frustration in certain areas of life is that we are just not making progress. This is why I am confident my formula will change everything for you. It is truly a game-changer.

Are you ready?
Let us begin.

The first thing about the success formula is that it is critically important that you know your ultimate outcome. Consider your health, for example. If your goal is to lose weight, you must be clear as to exactly how much weight you want to lose—this is your outcome. Your outcome must be crystal clear. It cannot be blurry. Clarity is power.

If your goal is to make money, you must be specific as to exactly how much money you would like to make—this is your outcome. Your target must be precise.

If your goal is to find a mate, you must decide upon your preferences. Consider the personality traits of the person you would feel comfortable dating or marrying—this is your outcome. This is important because if you don't know your outcome, you may be tempted to settle for anyone. Remember—clarity is power; you must know your outcome to be successful. There are no exceptions to this rule.

Whether your goal is about family, relationships, or money, you have to know your outcome, and specifically what you want.

The second thing is to take massive steps towards achieving your outcome. This is the fun part. Notice how I did not say to take baby steps—I meant to take massive steps. If your outcome is to go from earning $20,000 to $100,000 a year in income, you must initiate a massive amount of action towards achieving your goal, according to my formula. This is important—you must take massive action. The more action you take towards achieving your goal, the sooner you will achieve the goal. This is critically important for you to understand. When you take massive action, you reach your goal much faster because you put more energy and effort into it. This applies to everything.

I took massive action when dating as a young man. You can probably imagine what that means. You will never be ashamed if you take massive action instead of taking just a little bit of action. Not only does this formula work in every area in your life, but it works every single time. All you need is the courage to take massive action towards achieving what you want. If you want something to happen right away, you have to be courageous and take massive action.

Number three is to always have a coach. What this means is that when you have a goal in mind, you should find someone who has already accomplished that goal. Let me explain why having a coach is important. Every single one of us has said, "I wish I had known then what I know now," at least once in our lives. You have probably also heard that hindsight is 20/20. With experience comes wisdom. Wisdom comes from banging your head against the wall as you make mistake after mistake until you eventually figure the problem out and become wiser. It also comes from learning from someone else's mistakes. This means that when you are trying to accomplish any goal, it is always wiser for you to learn how to achieve it from a coach who has already accomplished what you are trying to do.

I have a coach in every area of my life. Be it spiritual, health, money-making, or saving and investing, I have a mentor no matter what I do in life because time is the only thing in life that you cannot get back. You can make money and lose money, you can lose weight and gain weight, but the one thing you can't get back is time, so you must value your time, and to do that, it is smarter and wiser if you always have a coach.

Consider business, for a second, and the goal of making more money for what you're doing. Find a good coach. This does not ring true only for business. It can be for any area in life including spirituality, relationships, and health. To find a good coach, go online and look up the type of coach for which you are looking. Make a note of the top ten or 20 in the field, and find out who they are—this is secret number one. Secret number two is to find out how to get in contact with prospective coaches. When you do, ask them a single question: what do I have to do to make it worth your while to coach me on what I want to know? These two simple steps are how you can attract any kind of coach you want.

The question from secret number two is powerful because when you find the coach, you're not asking him/her to give you something for nothing. You are telling the coach that you are willing to work for what they know, exchanging your work for his/her knowledge. It shows the coach that you don't want something for nothing, which is very attractive to a coach and says a lot about you. Once again, it's really simple to find a coach. Look them up online—the top 10 or 20 in that profession. Number two, get a hold of them and ask them what it is that you would have to do to make it worth their while for them to teach you what they know. This is step number three. Always, always, always have a coach.

Finally, step number four of the ultimate success formula: gauge your approach. Let's say you have the first three right—if step number four is not working, then change your approach. You need to have an acute sense to know if your actions are bringing you closer to what you want, and if things are not working for you.

Now that you have the four steps of the Ultimate Success formula, I want to share all of the areas of my life in which I use this formula to be successful. This is the most important formula of my entire life, so let us take a look.

In the area of health and fitness, I have used this formula, knowing my outcome. For example, right now, I am in the process of losing 23 pounds in 30 days. I am 223 pounds right now, and I want to be 200 pounds within 30 days—I know my outcome.

Step 2: I'm taking massive action toward my goal. What does that mean? As an example, I prepare the meals that I eat every day. I work out with a trainer using a specific type of workout that will help me to reach my goal.

Step three is critical: I have a coach. As an example, I am learning from someone who has already lost 25 pounds in 30 days, and I am doing exactly what he did. I am eating the same foods and doing the same workouts.

Finally, step four—let us just say if I am 25 days into the program and I am eating right, and I am working out, but I am not making any progress towards my goal of losing 23 pounds in 30 days. That is when I would change my approach, according to step four.

I use this formula in every area of my life.

Let us talk about making money.

Years ago, in December of 2016, I was actually broke. I had made some very poor choices with my money and found myself in a position where I was broke. I believe that if something is broken, you should fix it, and I used the formula once again.

I knew my outcome: to save $1,000,000 over 12 months—not make $1,000,000 but save $1,000,000 over 12 months.

Step one: I knew my outcome.

Step two: I took massive action. I got involved with a business, and I went to work like you would not believe. I worked harder in that business in the first 30 days than I have ever worked in my life.

Step three: I made sure I had a coach, someone who has already achieved this goal before, and I made sure I did exactly what they told me to do.

Step four: if I had not seen progress by following the formula, I would have changed my approach. I continually gauged my results to see if I was getting closer to my outcome. Guess what? It worked!

On December 1, 2016, I was broke, and by December 7, of 2017, not only had I made millions of dollars, but I had saved over $1,000,000 liquid in my personal bank account for the very first time in my life, and I did it over 12 months. I made copies of it, had it framed, and posted it all over my house. My point is that it works in every area of life.

Let me end by asking you this: how many chances do you give a newborn baby to walk? The answer is: as many times as it takes. Some babies walk in ten months, and some of them walk much later in life, but by the time that baby is seven-years-old, all babies walk. This is why I say to gauge your results. If something is not working, change your approach and gauge your results. If it is still not working, change your approach again and gauge your results. If it is still not working, change your approach and gauge your results. These four steps will work in every area of your life. They are very simple—extremely simple! Do it now, and I promise you that your life will instantly improve.

I'm Romacio Fulcher, The California Kid.

BIOGRAPHY

Romacio Fulcher is a well-respected international leader, trainer, coach, and much sought-after mentor. He is a highly successful entrepreneur. He tried the traditional college path and quickly found that it wasn't for him. Instead, he began as a cold caller, getting leads for a mortgage company. Under the owner's mentorship, he mastered the business, ultimately going on to start his own mortgage and real estate company to become a self-made millionaire by the age of 25. He was introduced to his MLM mentor and the industry of network marketing about ten years ago. He applied what he learned once again to become a top leader and earner. Although Romacio has earned millions of dollars, his heart is in training and inspiring others to transform their lives by helping others earn supplemental income along the way.

Contact Information
Facebook: https://www.facebook.com/romacio.fulcher
Instagram: https://www.instagram.com/romaciofulcher/
YouTube: https://www.youtube.com/channel/UC1zB-1jmoOpJE_hg_aEPNZQ

CHAPTER 23

SHADES OF GREY
By Samantha Jung-Fielding

"For someone so intelligent, what on earth was I thinking? How did it get to this?" These thoughts tumbled around my mind as I gingerly touched my scalp, feeling the coin-sized bald patches where my hair was ripped out.

Something did not add up. Why did I stay in an abusive relationship with a man who tried to murder me four times? How much more would it take for me to wake up and smell the coffee? As things turned out, only one single well-aimed punch, not leveled at me this time. Instead, he hit the dog. He struck so hard I heard its skull crack. At that precise moment, I finally understood that unless I acted quickly, one of us would die.

It was January 1996, and I was 28 years old. One week after that incident, I ran away taking only the clothes on my back. I drove out of Belgium and down into France, heading for the ferry back to the UK. At a safe distance, I phoned him. His voice raged down the line, irrational, coercive, and demanding: "Did you take the TV? When I get hold of you, I'm going to rearrange your face. You'd better get yourself home right now and sort this out."

Although I initially felt I had the upper hand, over the next nine months, I desperately wanted to go back every single day. It was impossible to make a simple decision on my own. The psychologist outlined how the abuser and I had become co-dependent.

I was floored. Me? Co-dependent? What do you mean? My mother has dependency issues. Not me! I'm Miss Independent, the most reliable one. I'm the person the family depends on to hide family secrets, and being the eldest child, I was there to always save the day.

Except I didn't save this day. And now, locked in with my abuser, I had to find the key.

So, how did it get to this? What exactly had I been thinking? I was the first person in my family to go to university, yet how useful is intelligence if you still make stupid life decisions?

It was a revelation when I discovered how the mind works. Simply put, it's a game of two halves— the conscious and the unconscious! Everybody instantly relates to the conscious mind because its job is to think and reflect, to make logical and rational decisions. In university, my conscious mind was continually stimulated, daily contemplating new ideas. This is why the conscious is also called the thinking mind.

However, did you know the conscious only represents about 10% of the mind's total capacity? The remaining 90% lies in the unconscious, whose main function is storage and processing. The unconscious is responsible for automatic body functions (like your heart beating, digesting food, and correctly positioning your limbs for sitting, walking, or even climbing the stairs). The unconscious carefully files everything you ever learned or experienced, plus it stores your values, beliefs, emotions, coping mechanisms, routines, and habits. The unconscious is also known as the feeling mind.

In this marriage of the minds, it's critical to note your unconscious is operational from the moment of conception, while your conscious mind first awakens around your 7th birthday.

As a result, during the formative years, small children absorb everything without question. Their surroundings are readily accepted as a baseline or "normal" reference point, and they are heavily influenced by the people around them. In short, without any filters, they act like sponges.

This is all the more alarming when you realize your early patterns set a foundation for the rest of your life. This means your brain was programmed by a toddler, who mostly copied examples handed on by past generations (i.e., your parents, or even older!). Of course, you had no idea things could happen any other way. So, when the controls were passed to you, you left everything in place, permitting those automatic and established patterns to run their natural course.

And the fly in the ointment? Unproductive historical family blueprints. This explains why in January 1996, I was repeating patterns from my mother's personal history. It's unlikely to surprise you that my mum also survived an abusive relationship, including a murder attempt when I was seven years old.

To recap, intelligence (or lack of it) was not a factor in my abuse. Nor did it matter what I had been consciously thinking. Because I was unconsciously following a pattern learned literally at my mother's knee. And, it got to this because I had not yet learned how to break this unproductive cycle.

Shocking! Unfortunately, the implied consequence was even more severe. Unless I acted quickly, I would subject my unborn children to the same fate.

So, where to next? Brazenly attempting to avoid my mother's alcohol dependency, I had already shifted from being independent to co-dependent. Moreover, it's not like she never warned against the pitfalls of extremity. As a child, I repeatedly heard: "You're so black and white. What about some shades of grey?"

Was I to infer grey is the color of inter-dependence? The very definition of black and white (or polar opposites) is that neither exists without its opposite. There cannot be yes without no, day without night, or up without down. Perhaps the crux of extremity is inter-dependence, and any spotlight must shine on the continuum between?

Unfortunately, I was stuck in extremity. Looking back, I noticed I was always out on a limb. Mum named my brothers in honor of their uncles. Yet, I was called Samantha—after the witch on television. In 1964, Elizabeth Montgomery took the lead role in Bewitched. A twitch of her nose launched a fashion for naming daughters, which reached me in 1967. Years later, I discovered the origin of my name is Aramaic, and it means "the listener."

I grew up believing I was magical with the unshakeable conviction I was put on this earth for remarkable feats. My formative years provided few examples,

Aristotle's wisdom set my course: "We are what we repeatedly do. Excellence, then, is not an act, but a habit."

According to Dictionary.com, "A habit is an acquired behavior pattern regularly followed until it has become almost involuntary." John Dryden said, "We first make our habits, and then our habits make us." Experience had already shown me that habits run the show, but I was also painfully aware of what goes on behind the scenes. Emotions (especially negative ones) muddy the waters.

My mum demonstrated how those with unresolved negative emotions (like anger, sadness, fear, hurt, or guilt) tend towards the systematic repetition of problematic behavior. Consider the use of stimulants (e.g., cigarettes, alcohol, coffee, sugar, recreational drugs), binge eating, video games, road rage, cyclical/abusive relationships, overthinking and procrastination. What mum didn't know was that getting rid of the underlying emotion can be enough to morph a behavioral issue into a productive pattern.

Emotional release is an immensely powerful and transforming process, which unhooks the negative feelings from past memories. This allows you to remember an event or person without re-experiencing the sensations that were previously associated with it. For years I blamed my mother for ruining my childhood with her drinking. By taking full responsibility for my own future and releasing this extreme perspective of my past, I chose to feel compassion for Mum. In doing so, I paved the way for a positive connection with my own daughter.

In his book "The Power of Habit," Charles Duhigg highlights how every habit is unconsciously designed to produce an emotional reward. Realistically-speaking, any effective and long-term behavior change must, therefore, begin with emotional release.

Unpicking enough formative patterns creates a wide, open space to accommodate your most eagerly anticipated emotional reward. When I abruptly quit my second marriage, nobody understood my choice. However, I was aware my carefully crafted happiness was lopsided, and I was gifted a moment of clarity. Propping it up further was establishing yet another dysfunctional dependency.

I have long cherished the idea of two equal and opposite people coming together to develop a wonderful continuum (or legacy) of well-rounded children. Over the years, I was gifted with three magnificent boys. I always dreamed of having a daughter, but becoming a single parent at 41 shattered this lingering hope.

Since marrying my third husband, I repeatedly joke that on good days we are equal, but on bad days we are very opposite. However, it was no joke when I fell pregnant just two weeks into our relationship. True to form, Mum was less than encouraging when I announced the news: "Oh God! Not another one!"

In authentic witch tradition, my daughter showed up 13 days late. Having already waited 19 years for this child, a few more nights of sleep were neither here nor there. Her entrance into this world was bold and breath-taking. She birthed unexpectedly at home, with a theatrical fall into the toilet.

I already knew her name would be Maya, after my mother's mother. Imagine my surprise when someone later told me Maya means "mother" in the Tupi language.

A few months before Maya turned 3, our family emigrated from the UK to New Zealand. Known to the children as "Yorkshire Nan," my mum was heartbroken. She was sure she would never see me again. Barely a year later, she had a stroke which left her paralyzed down the left side. Some months after, I flew back to the UK to surprise her for her 70th birthday. That was the last time she saw me. After that, she slowly deteriorated and passed away shortly before Maya turned 7.

While mum waited for her curtain to fall, I rang her daily. She could no longer speak, so the nurse held the receiver to her ear and told me Mum smiled all the way through our calls. I chose to remain in New Zealand for the funeral, and my best friend from childhood attended on my behalf. There is a significant time zone difference, so I had just jumped into the car with Maya when I saw a phone message flash up from my friend containing pictures of my mother's wake.

Apparently, "A Whiter Shade of Pale" was at the top of the charts when I was born, and Mum had requested this song be played at her funeral. Without warning, the first melancholy notes drifted from the car radio as I swiped through the photos on my phone. Unexpectedly, Mum's coffin flashed onto my screen, and I began bawling. From the seat behind, a small arm crept around my neck as my daughter asked: "What is it? Is it your mum? It's OK, mama. Just remember when you were a little girl with Yorkshire Nan."

Maya's simple words of comfort contained wisdom beyond her years. Deepak Chopra says that to break through ancestral bonds that inhibit your ability to create the life you deserve, you have to understand where they originate.

No doubt you know someone who has spent a lifetime seeking. Yet, the answer is right there before you. Your family blueprint is already cast in black and white. It's the shades of grey that seek your attention.

I implore you to tread with compassion as you release negative emotions and unpick your formative programming. Your very existence is a formidable gift. Take the opportunity to design your unique pattern of productive cycles and happiness habits.

And be mindful that your children will copy your examples (good and bad!). So, I strongly suggest weaving slings and arrows of outrageous fortune through your shades of grey. Therein lies real magic!

BIOGRAPHY

Samantha Jung-Fielding is an award-winning speaker, author, and performance specialist in the field of habit formation and behavior change. Master hypnotherapist, master NLP practitioner, and certified business mentor, Samantha believes unconscious brain and body patterns hold the ultimate secret to your success. Corporate burnout led Samantha to establish a charity, co-found a business networking organization, and develop a radio chat show. She also launched diverse businesses, initially in traditional fields before expanding into network marketing to deepen her reach. When her family immigrated to New Zealand in 2012, Samantha and her husband opted to raise their children on a small alpaca farm in South Auckland. Had they understood the alpaca predilection for mischief, they might have thought twice about this choice.

Contact Information
Facebook: https://www.facebook.com/samjungf
Instagram: https://www.instagram.com/sam_jungf/
Website: http://www.happinessence.co.nz

CHAPTER 24

THE IMMACULATE JOURNEY OF A FAITHFUL WARRIOR

By Serah W. Muiruri

As I walk on my journey, my view of the world is one comprised of many dots, intrinsically connected and unique in their formation, the world would be incomplete if these dots were not strategically positioned to complete the whole. I am amazed at the purpose of each dot, each of them leading to my final destination; without any of these dots, that destination would not be as intended. Often, the purpose of the dot is not revealed on the surface, and if it is, its significance does not fall into place until the end approaches. As I toy with these thoughts, I realize that life is an entity forever unfolding, one that is full of contrast, complexity, uncertainty, confusion, mystery, and above all, vulnerability. These attributes continuously shape who we are day by day, season by season, and year by year. It is a maze that never ends, from its inception and our grand entrance into the world until the time we take our last breaths at the end of our incredible journeys and onto the next one, a rebirth into that unknown world we get to experience beyond our time here, on earth. We are, for today. When our mission is accomplished, we are gone, a little while later. During it all, the world continues to evolve in ways unimaginable. Our world is quite different from that of our ancestors. What determines our existence? Our core beings? Our essences? Is this the philosophical dilemma of nature or nurture? I can't help but think that how we end is shaped by who we are as unique beings, our genetic composition, our upbringing, our cultural backgrounds, our spirituality, our metamorphoses based on our individual experiences, and above all, the choices we make along the way. These choices are critical in crafting the masterpiece of our valuable selves. When these choices are in-line with our core values, we are more fulfilled, and consequently, a healthy equilibrium is born.

As I compile these words, my desire is to speak to someone out there who is struggling on her journey in a world that is often quite unforgiving and let her know that the world can and is still a beautiful place, in spite of the dark clouds. No matter what you face at the moment, the sunrise is always visible on the horizon. My mission and ultimate desire is to spread hope to the hopeless, to plant a seed of joy in those at their lowest point in life, and to let those in despair know that it is in those desperate moments that a new "us" comes to be. Greatness is derived from atrocities. These critical moments propel us toward something more significant, that promises a greater tomorrow. If we allow ourselves to have that experience, it will prove better than the day before.

Here's an extract from a blog post I wrote on March 12, 2014, titled "Uncovering Our Hidden Treasures – Our Childhood Dreams."

All of us had childhood dreams. At some point, we all wondered what we were going to be when we grew up. Then, we woke up one morning to discover that our childhood dreams and hidden treasures had been lost along the way. How did we get here so fast? Where did the time go? All of us inevitably get caught in the proverbial rat race, speeding through life, entangled in our desire to sustain ourselves. Then, slowly but surely, time creeps away only to discover that our childhood dreams have been forever shattered. Our childhood dreams reflect the joys of yesterday and our hopes for the future. We should all protect these dreams, feed them, and encourage them to grow. We must take care of our dreams, for someday they may take care of us.

I believe we are all destined to craft our paths as we nurture our unique individualities, which only we can define. This uniqueness is what separates us from others. Our goal in life should be to walk along this path without losing sight of who we are as individuals. It is critical to adopt a mindset that allows us to live lives of purpose, no matter the situation. We are drawn to be clinicians in the human services field because we want to give back to society, be it giving time or talent, always guided by the philosophy that we are our brothers' keepers. We are great at serving others, but it is easy to get lost doing the very thing we love the most. As my brother's keeper, I challenge everyone to do a little soul searching this season—take a brisk walk, feed the birds, listen to your favorite music, do some gardening, or share a cup of coffee, tea, or a meal with a friend you haven't seen in years. Do whatever appeals to you. Let us revisit our childhood dreams to renew them. They say that it is in the giving that we receive, but to receive, we must learn how to give to ourselves first so we can give more to others. We have all been given what we need; it is time to rediscover our purposes in life, which will ultimately help to shape our destinies.

Let's dig up our hidden treasures and childhood dreams.

Blog Reflection:

I wrote this blog about five years ago upon my CEO's request, which was based on taking a reflective journey through my life. At that moment, I realized there were common themes in my life that were instrumental in shaping me into who and what I am today. My being is a product of many things, and most are referenced in my abstract. Vulnerability has been a constant in my life as far back as I can remember. I joined the Mary Leakey Girls School for my A-levels in 1988 at the age of 17. I was only going to spend two years at the boarding school before I hoped to enroll in a local University in Nairobi, Kenya, upon graduation. It was the very first time I would be leaving the nuclear family of my parents and siblings. I was young, meek, lonely, shy, and unsure. I was consumed by these feelings as a youngster, at it eventually had an impact on my academics. Back in the day, Kenya only had four fully-fledged universities, and only the cream of the crop could attend these institutions. I certainly knew I had the brains, but my struggles hindered my desire to be the best I could. I barely survived, made few friends, and basically became a lost introvert in the large crowd. My only goal was to get through those two years, and I would be miraculously fine.

Two years later, I was back in the village, having been left behind while my colleagues headed to those higher institutions. How could someone like me, someone who was "First Division" material, be in this vulnerable position. I was 19 before I realized I needed to gather the courage to face the world head-on. A spark of hope was born within me, which was ultimately a defining moment. I will be forever grateful to my parents and the rest of my family for believing in me and instilling the value of hope, which would be incredibly useful in my journey of self-discovery. I would soon become a high school teacher—a good milestone—and that was only the beginning. After a long struggle and many challenges, my parents found a way to send me to the US for further studies. That was the turning point in embarking on my journey of discovery.

It was a hot summer's day in August 1992 when I left Kenya for the US. It was my first time on a plane, and it was a long flight. I landed in Boston, MA, with mixed emotions. As I navigated my way in that foreign land, I realized that my life was bigger than me, and my experiences at MLGS had not only prepared me for good grades but something bigger. These experiences gave me the tools and ammunition I needed to succeed in the US in the absence of family ties, and with only a few friends with the same life experience as me. From that point on, I knew I needed to craft my life not only to realize my dreams but to help others realize theirs as well, my family included. As the firstborn in an African family, it was expected that I would share the responsibility of helping raise my siblings, but now that I was in the "land of milk and honey," my responsibilities extended to those in my community.

Expanding Responsibilities

As a Catholic girl who almost joined a convent, the life I have lived has been in-line with what I perceive to be my purpose. My faith in God has been instrumental in this, and without it, I fully acknowledge that I would not be who I am today. Being the first one in my family to travel abroad—and nearly the first young woman from my village (in rural Ngarariga, Limuru, Kenya) to leave Kenya to further her studies—I became a key player in assisting families with young adults interested in joining me in the US. It was a task I executed willingly, as it is what I was destined to do. Yes, I made many sacrifices along the way. Often, I could not see how I could continue, given that I was earning meager wages and attending college while assisting others do the same. My parents could not pay for tuition fees because it was an impossible undertaking, coupled with the fact that I was perceived to be a responsible adult at that point.

To remain purposeful became my goal, and I focused on it. Today, my actions have had a positive impact on many families. By allowing myself to be a selfless instrument of giving, dreams were born, and they continue to be fulfilled. Most of all, my siblings were able to add to my pool of successful stories, and I now have the joy of being a part of their successes, as well as that of the others who have crossed my path over the years. In doing this, I gained much more than I would have, had I focused solely on myself.

Marriage and divorce were important milestones in my life, and perhaps the very best example of life's complexities and conflicts. Of all of the vulnerabilities I have experienced, these are significant reference points for me. I was married in a happy ceremony on a lovely fall day on November 13, 1999, to a gentleman I thought was my best friend. We had, after all, been everything to each other for several years prior to the big day, and everything was just great. As fate would have it, my marriage crumbled ten years later. To say I was devastated would be an understatement, especially considering the added tragedies associated with the painful break-up. How could the thing I thought the most natural be that complicated? As a childless African woman, the thought of being divorced was beyond painful. How did I end up here? How or where did I go wrong? What did I have to show for it? Why me? These and endless other questions rang like annoying bells in my head, and a daily constant for many years following. My divorce was downright nasty, and the experience was unbearable, ranging from domestic violence, immigration threats, societal biases, and invasion of privacy, culminating with video from cameras placed in my dwelling and vehicles. I was followed and monitored every step of the way, making fear my sad reality. My soon-to-be-ex accused me of poisoning him when his health suffered a downturn. It was a baseless accusation, with no scientific evidence to back it up, a ridiculous allegation. The status of his health only served to prolong my agony, as I was forced to halt divorce proceedings until he got better. It was a hurtful, grossly painful, and trying time.

My world had changed, and now, my community doubted me. My only true confidants were immediate family members and a handful of close friends who numbered no more than three people at that point. I struggled to pick up the pieces, knowing I had no other choice. My faith in God kept me going, magnifying my resilience to greater heights. Amidst all of this, I focused on my job and caregiving responsibilities for "my Jen." Her being in my life gave me a sense of value and a higher purpose, keeping me grounded and focused. My coworkers at the time were a great source of support as well, and I remain forever indebted to all of them. It was at this lowest point in my life that I discovered an appreciation of and beauty in being alone as a journey of self-discovery. For the first time, I was able to distinguish between loneliness and being alone. My inner-strength was amplified; it was, indeed, the time for rebirth. Jen remains my purpose today, as do all of my nieces and nephews, who also give me a sense of validation. Nathan and Nayva have been especially instrumental in my healing process. I now know, for a fact, that my life is still unfolding. My parents, siblings, George, Martin, Grace, and Damaris, as well as my extended family, have continued to believe in me even when my strength appears shaken. I know my well-being matters to them, and particularly to my parents, who are aging gracefully.

Letting go of the bitterness was another facet allowing me to enjoy life after divorce as it was meant to be. It was not a simple decision; it was, however, necessary for me to move forward. Trusting that this milestone was part of my ongoing journey, I accepted the situation as it was. Thus far, I have gone through a restoration period, and by God's Grace, my success story is still unfolding. I know all too well that in faith, my destiny is well-crafted and spectacularly designed.

As a life coach, I understand that operating from a judgment-free zone is a happy place to be. Allowing myself to view others through their own lenses was another instrumental discovery. Most of the time, people do the best they can with the resources—which are often limited—they have; perfection is but a myth.

The lessons I learned:

- always allow yourself to be vulnerable, as it is in this state that optimal growth is attained
- humility is a virtue of honor—strive to be humble at all costs

- everyone crossing your path is there to fulfill a purpose, good, bad, or indifferent
- embarking on a journey of self-discovery when surrounded by adversity sets a beautiful stage for living a more fulfilled life, which often leads to the healing process
- nothing great comes easy, and if it does, the fruits are only short-lived
- always allow yourself to be an instrument of peace, regardless of the atrocities around you
- stay focused on your quest while searching for your true purpose in life—this revelation will save you in hard times and will provide a great reference point as you navigate through your day to day challenges
- once you find your purpose, strive to execute it at an optimal level
- become that which you were destined to be
- remain true to yourself and others; allow yourself to remain authentic to your values
- aim to have a positive influence on others, but do so with a sense of humility; expect nothing in return
- the good you do will always come back to you ten-fold
- where there is a will, there is always a way—just DO IT
- your inner-strength is greater than you realize—it is insurmountable
- the choices you make will have long-standing effects; choose wisely
- if you deviate from your intended path, forgive yourself and get back on track—nothing remains constant
- appreciate those around you
- have faith in a higher power (God for me)—this deity will always be a part of you and guide you through life
- spend time connecting to your inner-spirit—this regular check-in will speak to you in a gentle voice that sounds loud to your inner ear.

CONCLUSION

Trust that your journey will unfold and end the way it was meant to. Always guard your spirit and keep your head high, even when you have no clue what the next minute will bring. Seize the little moments in life, for they are gifts that continue to give. Your preparedness will influence your ability to receive the seeds that will grow into sweet fruit. Amidst this dance of life, we are but unique *dots* that make the world complete.

BIOGRAPHY

Serah Muiruri is currently licensed as a certified rehabilitation counselor serving as a master level clinician in the ever-progressive field of human services. Serah works for Nonotuck Resource Associates Inc., a local agency in the state of Massachusetts. Serah's career began in the early 90s after emigrating from Kenya, her country of origin, to the US. In her tenure of service, she has held several positions, serving children and adults with disabilities and often multiple diagnoses, including Autism Spectrum Disorders, intellectual and cognitive challenges, and acquired brain and trauma injury, among others. Her intrinsic drive to see others succeed has influenced Serah's attraction to the field of network marketing, and she has successfully created sizable networks. Additionally, Serah is a trained life coach, specializing in neurolinguistics.

Contact Information
Facebook: https://www.facebook.com/serah.muiruri

CHAPTER 25

RESCUE MISSION

By Sheen Marshall

I was on the last of my three jobs for the day. Driving around gives you plenty of time to think about yourself, your situation, and life in general. The previous two days, I had driven for twenty-seven out of the last thirty-two hours. It was brutal and dangerous because I was so tired that I was drifting while driving. I wasn't doing it because I had a love for driving. I was doing it to help make ends meet for my family.

During those long hours behind the wheel, the same thought kept going through my head: "What am I doing wrong?" Crying out to the Lord, as I have done countless times, I couldn't stop asking myself what exactly was I doing wrong that my life would have gotten to this? Life was nowhere near what I had envisioned it would be as a forty-year-old. My family was struggling financially, and we were just barely making ends meet. Our home was in foreclosure, and the business that I launched a few years prior was failing with $85k spent and not a single penny back. It was a terrible situation to be in, and one that I wouldn't want anybody to have to go through. I kept asking myself how a person with a B.S. and Master's Degree could find themselves living this way. Hadn't I done everything I was supposed to? I had followed the system of going to school, getting good grades, and getting a job. But this had led me to being forced to drive twenty-seven out of thirty-two hours just to earn enough money to, hopefully, cover the basic bills. There was that question again: What am I doing wrong?

Sometimes when you're drowning, the Lord will send you a life raft. It's incredible how often I see the life raft thrown to people and they push it away—and continue drowning. I refused to allow my story to end like that. I never wanted any handouts or freebies; I just wanted a shot. I needed the fire to burn inside me again. I needed to be reborn and to dream again. I needed hope. Fortunately, all of that happened for me in December of 2016. That was the day that an old friend that I hadn't seen in ten years flew into town, without telling me that he was coming, to throw me a life raft. Often, though, we will block or ignore a blessing because it's not wrapped up in the package that we want it to be or expect. So many of us become lost, meaning that we wander through life without a purpose. We haven't found the reason why we are here. I have been fortunate enough to find my purpose, and I'd like to go back a bit now to share with you how I discovered it.

As I look back to those grueling years of driving around as a delivery driver, getting no sleep for little to no money, I always knew that there was something bigger for me out there. I knew that this wasn't my final pit stop. I was still holding out hope that my failed business, a fantasy sports website, would take off, but the grim reality was setting in. I wondered if it was time to finally give up on that dream. But then what? That was the big question. I was previously a teacher for fourteen years, and I was certain that the profession was NOT for me. I disliked it so much that I let my teaching certificate expire because I didn't want it to be an option to fall back on. Leaving that profession, along with starting a business, led me to become a delivery driver for multiple companies. I remember showing up for work daily at one of my driving gigs and looking around at the others doing the same thing and wondering how we had all ended up there. There was a man formerly successful in banking, an armed forces veteran, a man who had a successful business and was now in his late seventies, as well as many others, including myself, and we were all forced to deliver delayed luggage for little to no money (that was one of my three delivery jobs). I kept wondering what mistakes we had made in life to end up having to do that for a living.

I used to be one of those people who believed that very successful people were always cheating or had some unfair edge over everyone else. Although this may be true SOME of the time, the fact is that very successful people spend lots of time working on themselves. How do they do this? It's called personal development. Amazingly, it's not taught in schools or society in general, NEARLY as much as it should be. When my old friend flew into town to throw me a life raft, what he did was introduce me to personal development. This was through the company I joined, which he was already a part of, that is BIG on personal development. As I started getting deeper into personal development and learning more, it was like an awakening took place within me. I started learning things about myself and others.

What's one of the biggest ways to work on personal development? By reading books, which was one of my BIGGEST weaknesses in life. I never liked to read books. Prior to February of 2020, I hadn't read a book in over fifteen years. As a man with two degrees, it was shameful to admit it. I have heard a saying in the past three years that goes, "success leaves clues."

As I started to observe and follow the people that had the life I wanted, I noticed that a common thing amongst them was the fact that they not only immersed themselves in personal development, but they also read books religiously. While I was sitting stagnant in my company and failing to advance in rank, I began to notice that, often, when someone was being interviewed, at the end of the interview they would be asked, "So what book(s) are you currently reading that you could recommend?" Person after person would quickly be able to call out a book on personal development that they were reading. And I sat there thinking to myself that I was going to beat the system and not follow the playbook by not doing any reading. Meanwhile, I'm looking dumber and dumber.

One of the crucial turning points for me was when I was fortunate enough to be invited to spend the day at the house of the founder of the company that I'm part of. It was on this day that I TRULY learned how much I DIDN'T know. The bible says, "My people are destroyed for lack of knowledge," and if that ain't the truth, then I don't know what is. The founder of my company talked to a group of us for six hours straight with no break. No matter WHAT the question or topic was, he pulled out a book and was able to flip to a page to answer or address it.

For six hours straight he did this, while I was sitting there saying to myself, "How stupid can you really be to think that the people who have the lives that you envy all read books constantly, but you think you're going to be the one that makes it to where they are without reading at all"? It was at that moment that I realized that my purpose in life was starting to take shape in a manner that I could only have dreamt about.

One of my favorite hip hop artists of all time is a rapper called The Game. He recently released his tenth studio album, which is titled *Born 2 Rap*. He was asked in an interview to explain the title of the album. His answer to the question has resonated so deeply with me that I recite it often. He said that (paraphrasing), "Everybody is born with a purpose. You don't know the point in your life that you are going to figure that out or when your purpose is going to find you." It took YEARS of struggling, headaches, sadness, and trials for me to find my purpose because it damn sure wasn't wrapped up in the package that I had expected it to be.

Nevertheless, I found it and embraced it. I was fortunate enough to discover my purpose in life and my destiny. The reason why this chapter is called "Rescue Mission" is because I must go back and rescue my people. When I say, "my people," I'm not just referring to the black community.

"My people" refers to all those who have realized that they weren't put on this earth to struggle to live paycheck to paycheck, be born and die in generational poverty, and not to advance personally or professionally. I believe that God didn't put us here to struggle. For so many, the daily struggle has become as routine as brushing your teeth and tying your shoes. My mission is to take all that I have learned on my personal development journey and use it to help those who haven't learned or been taught these things.

My people are crying out for help, and I am here for them. My mission is to show them that all of their dreams and desires can be attained by investing in themselves. You have LOTS of fight left in you to be the person that God has called you to be, but you must stay in the fight until you realize who you are and are willing to develop and grow into that person. Maybe you are someone who has yet to find their purpose in life? If that's you, then I say to you Godspeed because the journey will be WELL WORTH IT. Just keep fighting and it will happen. If you are reading this and have determined that your purpose is to rescue others as well, then our Rescue Mission has just begun, and I look forward to going into the trenches with you to help and save as many as we can.

BIOGRAPHY

Sheen Marshall is a network marketer who spent fourteen years as a school teacher. His career in network marketing has allowed him to help others find success in the industry and has also led him to discover his purpose, which is to help those who truly want to live and not just be alive. His passions include traveling, exercising, sports (especially fantasy sports), reading books, and listening to audios on personal development. He is also an active member of Omega Psi Phi Fraternity, Inc.

Contact Information
Facebook: https://www.facebook.com/rausheen.marshall
Instagram: https://www.instagram.com/big.sheen05/

CHAPTER 26

WHAT MOST PEOPLE DON'T UNDERSTAND ABOUT NETWORK MARKETING

By Steinar Pettersen

It's now more than two decades since I first became involved in network marketing. The first company that I was introduced to was selling a health product, and after attending the first meeting, I felt very excited about my prospects. I remember talking to my mother about it back then and telling her the things the company had told me during the meeting. It's so easy, I said, the product sells itself, and I'll earn a lot of money. However, I saw myself quitting my job after a couple of weeks because I thought the marketing opportunity would be such an easy and lucrative ride.

Readers who have worked in network marketing will recognize the kind of lies companies feed you when they're trying to enroll you—which I fell for—and the damage done by this sort of misrepresentation is something the network marketing industry must take responsibility for. It's the reason why many marketers all over the world have been overselling in a way that has ultimately harmed the whole industry and given it a bad reputation. The result is that new and innocent marketers struggle at the start of their careers.

During our discussion about the new role, my mum asked me straight out, "Who do you think is going to buy all these products you are selling?" "You are!" I said. "No, I'm not," she replied. She'd already noticed that the product had quite a small target market, and thought it was unlikely to have a mass appeal.

Looking back, I realized that my mum had already figured out the catch, but I had to figure it out the hard way.

Fueled by courage and enthusiasm, I felt I was moving forward, always looking for my next target, even though I hadn't enrolled in the company. Despite that, I was dedicated.

Around this time, a friend and his wife came to visit us at our home. I was excited as I started to tell our guests about the company and the product, and how I was expecting to make BIG money with almost no effort. But my friend interrupted me, saying, "Steinar, I've already enrolled with the company you're talking about." Surprised, I asked him how it was going. He told me that he had so far failed to enroll anybody or shift any product at all, despite having a basement full of products that he'd paid for. His wife suggested that he try to sell some of it, but he said she could have a go at doing it because he didn't want to. We all had a good laugh about that.

That proved to be a fateful conversation for me, and you could say that my good friend "saved" me from network marketing at that time. After learning about his experience, it dawned on me that this was far from an easy way to make lots of money with little effort. I think it's because of the way that the idea is sold to many freshly recruited marketers, as it was to me, that most of them don't understand why they end up struggling with their business and lose money. They're given an unrealistic idea of what's involved and think they've joined a network where they don't have to do anything but watch the money come rolling in. But guess what? Network marketing is hard work, like any other business. If you want to succeed in your career, you have to put in the hours.

Looking back, I think I have always been a dedicated guy, someone who easily gets excited about new things. I remember getting my first real permanent job at sixteen while I was still studying to become a mechanic. I vividly recall the day I drove out to the farm on my moped and applied for the job in person. It involved taking care of the animals, feeding them, gathering eggs, and milking the cows during the weekends and vacations. I was thrilled when I got the job and felt so proud when I was able to tell my parents. It was through that job that I learned to take responsibility at an early age. At that time, I was so dedicated to agriculture that I thought I was going to be a farmer. However, after finishing studying mechanics three years later, I already knew that the one thing I wasn't going to be was a farmer.

So, I got a job as a mechanic, working with pressured systems: first steam boilers, and then pressure air systems. Because I'm from Norway and come from the area known as the country's "Oil Capital," I had the opportunity to work offshore in the North Sea on oil installations as a technician on the air and firefighting systems. I had other missions, too, working as a course instructor when our company oversaw the gas contingency onboard. In total, I worked offshore for around fifteen years, an experience I value very highly. Then, about ten years ago, I moved into the sales department of the company. I had a flying start and made a new record for sales. Everything was good, and everybody was happy.

However, a few years later, there was a crisis in the oil industry, and everything came to a halt. Sales levels dropped dramatically, and my boss changed the rules about sales payments. Under the new regime, all sales personnel would only be paid for the sales they closed during personal sales meetings and not when orders came *through the house*. As I was selling the most expensive products, I was the biggest loser in the new system. I was unable to make anywhere near the same level of sales through meetings as before, so my income fell sharply, and before long, I found myself frozen out of the company.

You can imagine how my self-esteem plummeted. I was at rock-bottom. It was then that I decided to leave sales for good and never return. So, I got a job working as a technical manager in one of the biggest greenhouses in Norway.

Six months later, in January 2017, my upline Frank gave me a call out of the blue. He asked me if I would like to come to a meeting the following Sunday. I said I would if I didn't have to sell anything, and he assured me that it was just an opportunity that he wanted to offer me. So, I went along to the meeting and sat through the presentation. It all sounded so good that I got super-excited again, and agreed to sign on as a customer, although, in my heart, I really wanted to sign up as a marketer, as well. However, I remained somewhat skeptical, as I'd lost some money on other scams previously.

Consequently, I told Frank that I needed some time to consider signing up as a marketer but would stay in touch. I hit on the idea of enlisting some of my friends to join me at the next meeting, thinking that I'd wait to see their reaction; if they joined, so would I. If they didn't, then neither would I. I felt that this would act as a sort of insurance. But I made a big mistake; I revealed to my friends what the meeting was about.

In Norway, network marketing has a terrible reputation, and most people don't want to get involved. Norway is one of the wealthiest countries in the world, so people usually don't have to take a job if they don't want to. There's a welfare safety net for the unemployed, although it certainly won't make anyone wealthy. However, right now, like most people, I'm not sure how long this safety net will last after the coronavirus pandemic. I've heard people in Norway saying on the news that they've only got a few kroner in their bank accounts because the new payment system designed by the government to support people during the pandemic isn't yet working as planned.

Anyway, after I told my friends what the meeting was about, they bailed out, and I did the same. Some days later, my upline called me again and said that he'd heard that the meeting was canceled. I confirmed that it was and told him that I was still skeptical. But Frank didn't take no for an answer, saying that he was sure I'd regret it if I didn't enroll. I didn't like what he said but ended up enrolling anyway, and also enrolled one of my friends who was willing to take a chance.

Once I'd enrolled and started working, my old dedication for network marketing returned, and I found that I really liked the new company. Nevertheless, I remember finding many excuses not to pick up the phone and start recruiting new people. At the time, I didn't have the right skills to do it successfully.

It's worth mentioning here that, in my experience, this is also something people starting out don't fully understand about network marketing—to do the work successfully, you need the proper training and the tools.

I soon started to join new meetings and began taking part in training sessions. I remember we had a visit from one of the company presidents during a meeting. He asked the room, who was willing to leave their comfort zone. I was the first to raise my hand, and he asked me to come up on the stage. I remember him telling me about some of the events that were planned. It was profoundly exciting, but he didn't understand why I thought all this was such a big deal.

Not long after that, I attended my first event in Torino, Italy. It was a fantastic experience, with a far bigger crowd in attendance than I had ever imagined possible. Eric Worre was conducting the training, and I could relate to everything he said. As Eric was lifting us all up with his step training, I was so fired up that I dared myself into doing something I would never usually do. During a break in the training session, I was chatting to a musician from a band that I like, when Worre started walking through the audience. I thought to myself, if he passes me, I'm going to give him a handshake. And that's precisely what happened. Later, Frank said that I was the only person who had successfully managed to divert Eric from his usual routine at such events.

I must say that it was at this training event that I think Eric Worre changed my life. I really want to thank you for that, Eric!

After the event, my career suddenly kickstarted—I went from Level 2 to Level 5 within a month. I remember being so full of energy that I sometimes had to let off steam by screaming in my car as I was driving along. Seeing my success, the company invited me to a leadership conference in Vienna, where I met the founder of the company.

My motivation to work was at its peak, plus I had great mentors who helped me get better at the job. We all put a lot of time into it. That reminds me to thank my lovely wife, Siv, who patiently put up with me spending so much time on developing the business. From time to time, she joked that the new business was my new wife.

Even when Siv and I were on a holiday in Spain, Eric Worre was with me on the sunbed, not physically, but in my ear and on the screen. I was learning how to recruit twenty people in thirty days. While I was training, I also started to figure out how I could put my own twist on the job. I began preparing for presentations, calling my father from Spain, and instructing him how to do a drawing for me that would make it easier to get my message across to people at meetings. By the end of the holiday, I had completed the training, which was like getting a monkey off my back. I found I was genuinely enjoying the business and having fun.

Many people don't understand that an important aspect of network marketing is the support you get from others. Plus, you can start your own business from home, and no formal educational qualifications are needed if you're willing to work hard. As I see it, one of the biggest problems facing the industry is that many people aren't ready for that level of responsibility and hard work. Several factors can get in their way, and the biggest is the fear of rejection. Many starting out are simply not adequately prepared for this, so when the fifth person says no, they give up. There's an easy remedy to this problem: if uplines give newly enrolled people more time to do the training step-by-step, then I'm convinced that many more would succeed. It seems likely to me that, after the coronavirus is contained, people will be more receptive to the business and the network marketing concept.

Sadly, the network marketing business I worked for closed in Norway about two years ago. It gave a lot of people an excuse to quit. I didn't quit. Instead, I focused on getting more training. In fact, during the family's celebration of my mum's seventieth birthday in Italy, I was still doing live training courses with Frazer Brookes on my cell phone. Thanks, Frazer, for all the great training sessions, and the books!

I'd like to send a special thank you to Matt Morris, who has supported me daily with great emails, live training, and other videos. He's always happy and full of energy, and I want to thank him for all the times he's lifted me up when the going was tough, and I wasn't producing any results.

I'll sign off by saying, because of network marketing, I am now back in sales.

BIOGRAPHY

Steinar Pettersen comes from Sandnes in Norway. He is an entrepreneur and business coach. After starting out as a mechanical engineer working offshore in the North Sea for the oil industry, he has since enjoyed a long and successful career in network marketing. He has participated in many network marketing training events hosted by industry luminaries such as Eric Worre, Tony Robbins, Frazer Brookes, and Matt Morris. Steinar is also the founder of a video production company. He's now dedicated to helping people succeed in their network marketing business.

Contact Information
Facebook: https://www.facebook.com/steinar.pettersen.98
Instagram: https://www.instagram.com/steinarsin/
Website: https://www.futurehomeoffice.com/

CHAPTER 27

FROM GO TO WOE . . . AND BACK AGAIN

By Stephen Davis

I was born at the bottom of the world with sixty-six million sheep and three million people, and pastures and scenery as far as the eye could see—my home, New Zealand.

A beautiful place and at the time when parents left their kids by themselves, cars were packed like sardines, and you left home without locking the doors . . . well, scratch that last bit, I grew up in one of the lowest economic places in New Zealand, and if you didn't lock the door, then you would probably lose that door among other things.

Devoid of any cell phones, internet, or video games—in fact, there was barely a TV with two channels—we made our own fun, created our own adventures.

My childhood was full of life and mischievousness. I belonged to a generation of kids who were left to their own devices for hours on end and, as you may imagine, did everything you can think of.

We turned the house into a bomb zone, the coffee table into a sled, we even practiced flying off the roof, jumping off rooftops and through windows of numerous houses, played tag around glass doors, somersaulted over concrete fences, let strange kids in our homes, lit matches, and firecrackers inside, made our neighborhood block into our "Olympics" park, turned our park trees into makeshift climbing walls, and crawled underground into partly-laid pipes by local authorities with nothing but a torch and courage.

I broke things, lost things, messed up things, climbed everything you can imagine while safeguarding my explorer spirit.

It was not surprising that I ended up with battle scars and in trouble several times. But I wouldn't trade it for anything, for it made my childhood memorable and led to an adventurous life.

I developed a knack for losing things. I lost my school term bus pass regularly and ended up running to and from school daily. I became extra fit, but it wasn't so great when the bus driver who often gave me free rides, asked my mother why I was running all the time? Ouch!

I was also "famous" for repeating the same mistakes. How many times I spent kneeling on the lounge floor with my siblings, saying sorry to mum and promising never to repeat the behavior, only to do it all over again the next week. The perils of such cyclic behavior came to haunt me later.

I harbored a few insecurities. Growing up with mixed heritage, I didn't fit into either culture and there was prejudice too. My parents constantly fought, and this meant that from as young as a preschooler, I was often the barrier between whether one got hurt or not.

We moved cities at a key time, which meant that I had to start new friendships and adapt while tackling the adolescent worries of pimples, hair growth, and low self-esteem issues.

In my new city, on New Zealand's beautiful South Island, my high school science teacher Mrs. Metcalfe, who upon seeing an insecure young man, said, "Stephen, you have a beautiful smile." To date, I am grateful to her for the positive reinforcement she gave me.

Sometimes the simplest comments can leave the most profound impact; positive or negative. Since then, most people often notice and compliment my smile.

Being "famous" as a child can be far more fun than painful. But as an adult, it entails entirely different things.

Remember when I said I was known to repeat the same mistakes?

I led my life by following meticulously carved plans. I was striving to fulfill my vision. I was married, owned a house, had a growing family, and a career to see me through to retirement. I had money saved and money growing, all was going according to the plan.

I was leading a life that most people dream of.

But as fate would have it, in a short amount of time, I lost it all. I came home from work one day to find the house half-empty; my wife of ten years was gone with the kids. I was forced to sell the house and watched half a million dollars disappear down the drain. With my career finished, my reputation hit rock bottom. Some "friends" and "loved ones" turned a blind eye while muttering "God bless," "All the best," etc.

I felt that the Universe was out to get me and I crumbled under surmounting pressure. Blaming God, the Universe, everyone, and everything else, I retreated deep into my shell. My health plunged, my hair greyed, my weight ballooned, and for the first time in my adult life, the future looked bleak.

The pain was so great, the depression and numbness so debilitating that I struggled to even leave my parents' home, which I had returned to for some respite.

I was barely alive, let alone doing anything else. My life had fallen apart dramatically.

But, lost in the middle of this perfect storm, I found a way through. And no matter what you encounter, I assure you there will always be a way through for you too.

Here's what I discovered, and may it help you in your time of need and guide you toward the life that you seek.

Key 1: Take Time Before You Are Forced to Take the Time

I remember the night that was the culmination of everything that occurred. I had no home, not even my parents'. Out under the stars at the beach, I contemplated my life and the possible future that lay ahead of me.

I had not pondered so deep in decades or asked any existential questions. It's no wonder I was a mess.

It is better to take a deep, hard look at your life before you have no other choice than to do so.

Key 2: Reach Out; Save Yourself Unnecessary Pain

At times, a third person and their objective perspective could become a key difference between a life mended or one torn asunder.

Before I slept, I phoned some people who played heartfelt roles in my life, and in doing so, it stopped me from doing anything stupid; stupidity which would have further derailed my life.

Key 3: You Can Change

Being on my own, with nothing, going nowhere, I had plenty of time to think. During that time, I had a deep realization.

It didn't matter wrong had been done to me, and how bad the situation is, it did not have to dictate the rest of my life. I could change it; the power was in my hands.

The realization freed me, and I began to let go. I was free to rebuild anew.

For the first time in ever so long, I slept peacefully.

Key 4: Look and Listen Carefully for Life's Clues; Learn early, Save Pain Later

Life has a way of sorting us all out—call it God, a higher power, or the Universe.

I started recognizing patterns in my life; different people, places, and times but repeated situations, with an almost déjà vu precision.

I thought, "How could this keep happening?" "What was the connection?"

I then had an epiphany—these were situations that I hadn't learned from, situations I had not resolved properly. It hit me like a ton of bricks.

In hindsight, I could see that throughout my life, I was "famous" for repeating the same pattern of dealing with situations. Consequently, I hit the same insurmountable walls over and over again.

It was clear how each similar situation became increasingly worse, leading up to the hardest time in my life.

I created a recipe for disaster by putting a band-aid on deep wounds and carrying on, until my life fell apart.

Key 5: Your Life Is 100 Percent Your Responsibility

I gained a new perspective and was shaken out of a victim mentality— "Why me?" or "How unfair!" Instead, I replaced it with a new understanding that my life was my own responsibility, no one else's. Soon, the questions changed to, "Why not me?" or "Life can be unfair, so what?"

It was the kick in the proverbial backside I needed, and I was so grateful to finally learn the lesson.

You can let life do you in, or you can take life and make it happen for you.

Key 6: Clear the Rubble of Your Life

It's one thing to get through something, it's a whole another to get over it. Throughout my life, I recall many situations where I held grudges for such long times. It was during the hardest time of my life that I felt the full weight of holding onto those grudges. I could feel the pull of those grudges dragging me down.

I lived for two years in such a damaged way that it compelled me to think of the finiteness of life. I came out, and adopted the motto, "One life, live it fully, have no regrets."

I properly processed what happened, felt the pain, forgave, let go, and moved on. Notice how I didn't say hide or forget the pain. If you try and do this, the pain has a nasty habit of resurfacing again, in unattractive ways, and at embarrassing moments.

I decided never would I let any situation ever cause me to dwell on it so long again and I haven't.

Ask yourself would you rather spend fifty years or more getting over something, blaming others, and being right . . . or let it go and live your best life?

Time does not care if you mope around for five minutes or fifty years. It ticks on, indifferently.

Now unless you know that you have an extra life or two up your sleeve . . . save yourself unnecessary life-sapping energy.

Key 7: Step Forward, Do Something Constructive with Your Life

From the worst soil grew a bud of hope. Within a week, I had a roof over my head, my kids with me, and shortly after, my new employment secured. I then began to rebuild the shambles of my life with a new foundation.

I took counseling, gained inspiration online and offline, and connected with people globally. I watched, listened, learned and grew, spent time discovering who I was, and doing things I loved.

A lifetime of developing an unhealthy mindset takes some time to dismantle.

I did not shy away from asking difficult questions to myself and others, for I knew it was needed to overcome the hurdles of a regressive mindset. Find your inner child and be courageous to give new things a go. In doing so, you will stretch yourself, grow, and discover the whole new *you*.

Remember, no will always remain a no if you choose to do nothing. So ask, do, and give yourself a chance of being the best *you*.

FINAL THOUGHTS

Why do some harvest their true potential, find success, and be financially well off, while many others continue to struggle? It's the barrier in their mind that gets in the way.

So . . .
Key 1 - *Take time before you are forced to take the time*
Key 2 - *Reach out; save yourself unnecessary pain*
Key 3 - *You can change*
Key 4 - *Look and listen carefully for life's clues; learn early, save pain later*
Key 5 - *Life is 100 percent your responsibility*
Key 6 - *Clear the rubble of your life*
Key 7 - *Step forward, do something constructive with your life*

Be that person who considers things most won't, take action where most don't. . . and you'll gain the life that most miss out on.

And Where Am I Now?

Healthy, fit, in a loving relationship, a great dad with wonderful kids, an author, a successful businessman, and moving quickly toward fulfilling my lifetime vision of being a global mentor, trainer, and inspiration.

Like nothing will stop me, let nothing stop you.

Break preconceived ideas, notions, and a lifetime mindset. Let go of pain and the things of the past.

Remove the barriers, build afresh, and abundant success will follow.

BIOGRAPHY

Stephen Davis is a mentor, coach, and successful networker. He is passionate about helping others to succeed, inspire, and impact their lives. He works out of his home in the beautiful garden city of Christchurch, New Zealand. He likes spending his summers camping and adventuring locally and abroad with his darling and their four children. If you look closely, you might find him dancing among the treetops . . . but not too closely, he has already fallen once and broken both wrists.

Contact Information
Facebook: https://www.facebook.com/StephenJDavisOne/
Instagram: https://www.instagram.com/2dreambelievelive/

CHAPTER 28

JUST DECIDE - THERE IS NO OTHER OPTION

By Steve Eastin

What happens when you finally decide that failure and living an average life is not an option?

What's holding you back?

Is it your age? That's only a number.

Your ego? Do you think that you can do it on your own?

Your choices? Okay, maybe you haven't made the best of choices, but you have to let that go.

Your past? Let it go.

Your mistakes? We all make them; just let them go.

Your financial situation? There are a lot of ways to make money—take a risk once in a while.

The negative people around you? Find new people.

Your own negative thoughts? Start creating positive ones.

The haters? This is my favorite one—prove them wrong.

Easier said than done, right?

My journey to mastering the above started when I was 11 years old. My mom and dad were entrepreneurs at heart, and they worked extremely hard. I watched them go to work, buy an apartment complex, create businesses, and build relationships.

Around the 11th year of my life, my dad made a statement that has stuck with me ever since: "You will never get rich working for someone else." This indicates that proverbial turning point (a noun referring to a decisive change in situation, especially one with beneficial results).

Dad led by example; his actions said what words could not. In the book *Rich Dad Poor Dad* by Robert T. Kiyosaki, the Rich Dad describes my father to a T, but I did not read that book until later in life. It wasn't until reading the book and taking stock of what I know now that I realized that my dad and mom were actually mentors and teachers.

Dad and mom eventually fired their bosses and continued with their investments and businesses. They were true entrepreneurs. Unfortunately, my dad died in his semi-retirement at the age of 44 in a small plane crash in Alaska while enjoying a hunting expedition. I was 19 at the time. Dad was on his way to millionaire status, but Death and Destiny had different plans for him.

After watching my parents when I was 11-years-old, I decided to get a jump on things and start my own lawn mowing service. I made some contacts, lined up some customers, and away I went. The money was good until the weather turned cold. The money after that came in the form of payment for household chores. It was not as good, but there was still money coming in. I always had multiple jobs throughout high school, but nothing significant.

When I hit 20-years-old, a new turning point occurred when I married my high school sweetheart. I convinced her that I would give her everything she desired because I would be a millionaire by my 44th birthday. I promised that our lives would be remarkable, worthy of attention. Unfortunately, I never achieved my goal. My theory is that I started to work for other people, helping to make their dreams come true instead of my own.

I worked in a mailroom and as a bartender. I still mowed lawns, and I was a cashier at Target. There were also some purifier sales and a job as a sales representative later in life. This pattern continued, allowing me to support my wife and enjoy the birth of our daughter. We lived in luxury—kidding! It was a two-bedroom apartment where the cockroaches were so big that we named them and called them our pets. Through it all, I knew that if I kept working at these jobs, it would soon pay off.

I worked my way up from that mailroom to customer support; at least I was moving up. It still felt as if something was missing. I knew that I was not answering my calling, nor was I keeping the promise I had made to my wife. I had this burning desire to help people and be in a position of authority, so I quit pretty much all of the jobs I had, including the customer service position. Next, I decided to be a police officer at the age of 30—it was always something I had wanted to do.

Supporting two children now, I graduated from the police academy—from which, I might add, several people were pretty surprised that I had (take that, haters). I first became a reserve officer, then a paid dispatcher, and I was eventually sworn into the patrol division. Trust me—I could write an entire book on that 18-year tour of duty, but I'll save that for another day. I will, however, tell you that it changed my life.

Believe it or not, my pattern of multiple jobs continued during my police career. While working as an officer, I started a trucking company, a construction company, and a trailer refurbishing company, all of which were operating at the same time at one point. I managed employees, did the accounting, spent time at the various locations, swung a hammer—you name it, I did it all while performing my police duties to a high standard. I never slept, but by God, I was going to be a millionaire by the age of 44. My future was bright. The money was pouring in.

All was good until the economy crashed. I will spare you the details, but shit just happened. I lost everything in a blink of an eye: savings accounts, toys, businesses, and the house. And then the ultimate happened: bankruptcy.

If you have ever experienced the feeling of not knowing where you and your family are going to live after you have uprooted them from everything they knew, it cannot be explained, but it sucked!

We lost the home on our 60 acres, but we were blessed to find a townhome in our community that served its purpose over the next few years.

During that time, I chose not to quit but stayed more positive and more determined than ever to succeed.

The bankruptcy taught me that I was doing it all wrong. Though I was working extremely hard, I had decided to assume the responsibility and blame for my situation.

I truly felt that God had put me in that position to let me know that striving to be a millionaire was about nothing but greed. And something was still missing, but I could not figure out what it was.

I got my real estate license in 2009 while still a police officer, continuing the pattern of working multiple jobs, and it has worked out well over time. I was back in the financial fight, but unbeknownst to me, I was still in the proverbial rat race.

These two professions began to take their toll on me, becoming more demanding and time-consuming. There was no such thing as sleep, family time, or vacations. Unfortunately, I became more selfish than before, which led to a downhill spiral for the worse. I started to become a recluse, I did not want to spend time with anybody, and I just stopped being me. I was not the kind of person others wanted to be around.

One day while on patrol, I received a text message from my daughter. In it, she laid out the truth. She advised me that I could do anything, that she and the rest of the family wanted me back, and that I should retire. This was interesting because I was the type of person that you could not tell what to do, nor would I take someone else's advice. It was a brave thing my daughter had done. Thank God, I listened to what she said. It struck my soul, and I knew she was right because I felt at peace.

Ironically, my commander called me into the office later that day to see what I had decided about my schedule for the following year. The schedule had been written on a whiteboard in his office. I found myself staring at it long enough that my commander asked me for my decision once more. I knew I had to save my relationship with my family, and I had to fix me. I looked at my commander and told him I was giving him a week's notice. He stated, "You're f****n kidding me!"

After we determined that I wasn't kidding, I agreed to a two-week stay to assist with training my replacement. After the two weeks, my agency gave me a fantastic retirement party, and I left law-enforcement. I am very grateful for my agency, and I would not change anything about my law enforcement career. I believe it was a journey that I needed to take.

To all of the family members of men and women in law-enforcement, the military, and first responders: thank you so much for what you do, I know what is involved, you are special, and I pray for all of you every day.

Holy crap! Now, what would I do with no job, very little savings, a small 401(k), and no real estate deal in sight? I guess you could say that I took a leap of faith.

I was pretty much in survival mode for the next year or so. My savings were not big, and the IRS loved me. I searched for my next turning point and what I had been missing.

The answer was simple: I just dropped the ego and asked for help. By that time, I had surrounded myself with some amazing mentors who gave me the advice to read books, take pride in myself, invest in some self-development classes, my mind, my financial education, and my training, so I did. What do you know? They were right. Most importantly, I devoted myself to God and stopped doubting him and the power of prayer.

I finally understand that you can't always do things on your own and it is okay to ask for help. I've learned about financial strategies and that relationships are more important than anything. It taught me how to create a focus board and values list and to visualize my goals and create affirmations. Trust me on this one: write it down, or it won't happen.

I now know my destiny, and I have found my core purpose in life. Life is no longer about my wants, needs, or material possessions. It is about the wants and needs of other people. It is about helping people to become leaders. It is about helping people succeed in helping others invest in themselves. It is about helping people to create a lifestyle of their own that they can control. The bottom line is, I needed to figure out how to become a leader for the right reasons before I could help others do the same.

To all who are reading this book, decide that failure and living an average life is not an option, and age is just a number. Start your positive transformation today. You can do it. I believe in you.

To my wife: I have not quite reached millionaire status yet, but thank you for being so remarkable and never giving up on me, no matter what.

BIOGRAPHY

Steve Eastin has been in the public eye for over 20 years. He has worked with thousands of people in different aspects of their lives and is deemed by friends as a successful life coach. His educational experience consists of ethics training, crisis intervention, crisis and business negotiations, and leadership and self-development. Steve is a successful entrepreneur with growing businesses. He is married to Brenda since 1986 and is blessed with a son, Danny, a daughter, Amanda, a son-in-law, Calvin, and granddaughter, Indiana Cree.

Contact Information
Facebook: https://www.facebook.com/profile.php?id=100009492462937

CHAPTER 29

PERCEPTION IS EVERYTHING

By Syen Yap

Have you ever wondered why some people seem to have so much good *luck* in their lives with everything always going their way? They just appear so positive and bad things never seem to happen to them.

I'm embarrassed to admit it, but sometimes I want to hate them, or at least feel green with jealousy. It's not fair, right? Well, it's always been a mystery to me, so I decided to try to discover if some people are just born lucky. And, maybe, along the way, I might figure out how to get luck on my side.

I came from a normal family with four siblings. However, my parents got divorced when I was in secondary school—when I was about fifteen years old. A lot of people might think that might not be a great situation, especially during our growing-up journey.

But today, I'm actually glad about that event; we children no longer had to keep listening to the arguments or endure the hostile atmosphere at home every day, when we came back from school. I now think it's good that both my parents took the next step, so that the changes allowed them to live much better lives.

Yes, it was challenging for me and my siblings when our parents separated. I followed my mom, and my siblings followed my dad. I took this path because

I felt that I understood why my mom made the hard decision, and I wanted to at least be by her side. My siblings and I communicated by letters, with me only going back once a week during the weekends to visit them.

There was one time when my youngest brother forgot to bring the house key, so, after he came back from school at around one in the afternoon, he had no choice (without a phone) but to sit by the front of the house, under the hot sun, waiting for another sibling with a key to arrive around seven in the evening. After a couple of hours, the next-door neighbour realized the situation and allowed him to wait inside their house, while he waited for the other sibling to return later that evening.

When I learned this, I felt so sad that my younger brother had to go through that. However, there wasn't a single time I blamed my parents or the divorce for it. I decided back then to always believe that things happen for a reason.

I am proud to say that, having gone through this life-test while growing up, all my siblings became very independent and mature compared to their friends of the same age. We rarely asked for help from a parent unless it was absolutely necessary. We monitored our own education, deciding on the courses we would take by ourselves. Even then, all of use understood that each person has to craft the life they want by themselves—that we are all responsible for our futures and no one else.

Fast forward to my life as an adult, and you might agree with me that my difficult growing-up journey had a purpose behind it. My attitude about life back then has now come full circle, and I still make decisions on the basis of my belief that our responsibility in life is to be the type of person that is a blessing to others.

My first job was in the hotel industry as a sales executive, followed by a second in the exhibition industry. Within these two entirely different environments, I was blessed to have a supervisor who really took care of me, always fighting for me, and earning my goodwill. I discovered that, as with growing up, there are workplace politics that threaten to drag us down. However, I instinctively chose not to participate in them, simply because such entanglements create so many problems—and there is never any gain.

Here are a few examples: Once, I shared this practice of positive perception with one of my colleagues, Joseph. I told him that I thought I must be a truly blessed person because I had so many great people around me. He replied, "That's because you always have an attitude that everyone appreciates—you always look at everything in such a positive way. So, all the negatives seem of such little importance that you cease to care about them."

Then, a year ago, I became involved in my first network marketing company. At that time, our team was not very strong. It seemed as though many people were thinking that they would wait until the team grew stronger or the brand become popular before they would commit to joining because they thought it would then be much easier for them.

But a voice inside whispered to me, "Why don't we become the *pioneers* and make it work when others won't? Isn't it challenges like this that create the greatest opportunities for us to learn and grow as people? The best memories we have in life are often of overcoming the challenges that, at the time, felt like the most difficult. But I say that those are the times when there is the greatest opportunity to build a strong foundation.

And, although our team was small, I'm so grateful that we had a great leader with the long-term vision of building something strong together. How grateful I am to be to be part of such a team!

Then, in early 2020, the COVID-19 pandemic arrived—the virus that took away thousands and thousands of human lives. People everywhere are suffering because of the coronavirus. Countries are in lock-down, with people asked by their leaders to stay at home. The world has been forced to freeze all economic, social, and physical development activities. Many people are genuinely worried whether this will be the end of the world, as the impact of the virus seems greater than any other event in human history. And it's happening globally. Will we survive? And if we survive, how will the world function in the future?

Yet, a few months on, the great news is that there are many signs that Mother Earth is healing herself. Plus, there are other positive outcomes—like the decrease of air pollution from factories and autos. We have better air quality. And not only that, but the quality of water has shown significant improvement in Venice, as well as in rivers in India. Wild animals have appeared in groups to reclaim their territory, and the Earth's ozone layer is healing by itself.

But does this remind us of how much damage we have done to our Mother Earth for the sake of so-called progress and economic gain? Can the pandemic be seen as a warning, a lesson from Mother Earth to force us to stop and think deeply about what we have done with her kindness? Have we wondered why COVID-19 has spread so quickly and lethally amongst humanity, forcing us to stay at home for the safety of ourselves and others, while the animal kingdom remains untouched? You could say that the animals are now claiming their "freedom" back from us.

Maybe we should be paying greater respect to Mother Nature? As humans, we claim to be the most intelligent among all the living species on our planet yet when we look at the bigger picture, we see that we only play a small role in the greater scheme of things.

Perhaps Mother Nature is reminding us to be more than just takers, selfishly squandering her bountiful resources? By focusing only on our personal gain, have we truly forgotten that we should also give back and show our care for her?

Today, I sit quietly in appreciation of the situation. This pandemic may be the biggest opportunity for humankind to grow and gain that it has ever had. It may be saving us from the day of reckoning that will come at the end of the world. I believe that, if we wake up and take better care our planet, we might save ourselves from utter extinction.

If I may, I would love to share with you an idea of mine that comes from the center of my soul; today, globally, I believe there are two types of people facing this pandemic:

Type 1 – Focusing on the virus, blaming others as to why it is happening, while the virus continues spreading due to the stress created by peoples' constant complaining.

Type 2 – Accept the situation and take it as a challenge, looking for a solution while feeling grateful for being safe and healthy.

Please don't misunderstand me—I'm not asking you to be oblivious to the fact that bad things happen in life. Of course, they do. And it's okay to acknowledge that we do not know how the future will turn out. But my life has taught me that what matters is how you choose to look at things: Do you perceive certain situations, like the pandemic, as "Bad Luck" or "A Great Challenge"?

I wish to share with you the secret that has helped me in my life: If we decide to take up the power residing within us that comes from choosing how we respond to life's challenges, that directly affects how we feel inside. Instead of choosing to be a victim, we can choose to live as great students, constantly pushing ourselves to rise above the expectations of our great Professor and earn the best grade we can. I believe that power lays in the way we look at any given situation and our mindset affects the consequences resulting from the actions we choose to take.

It begins with your choice of not *what* but *how* you choose to believe. It's all a matter of perception. For instance, I choose to believe that things happen for a reason, and that outcomes are a result of what we did or failed to do in the past. Think about it this way—that life may be testing you, giving you an opportunity to begin to transform yourself into who you're meant to be— a better version of yourself!

Yes, I could have chosen to blame my parents about not giving us the best lifestyle during my childhood. Yes, I could choose to complain about why my career path has had so many ups and downs and difficulties. Yes, I could choose to blame others for this pandemic. But there is something that I know within the depths of my soul: None of these behaviours will serve me well. None of these beliefs will change the situation for the better, and all such negative emotions only hurt me.

But if I *choose* to accept the situation, switching my way of thinking—my perception—gives me the *courage* to face all these wonderful challenges with positivity. Because, in my experience, challenges usually mean an opportunity to grow and become a better version of myself.

Remember, as human beings, we do not have the power to control what will happen. But we do have the power to choose how we respond to life's challenges by deliberately aiming our intent towards the outcomes that we want to achieve.

My friend, please be grateful for whatever happens in life. Every event is a meaningful piece in the grand puzzle of life, one that helps us paint the story of our lives that we can recall again and again when telling it to our great grandchildren. When we inspire them with our courage to make the best of life's challenges, we are seeding the power of positive perception into their minds and ensuring that their world will be even better than ours.

BIOGRAPHY

Syen Yap is a professional executive in the exhibition and network marketing industries. Challenging career environments have built her tough and resilient character that always looks for solutions to problems through a positive mindset. Her life experience and positive attitude in life has successfully helped her create an influencing power over more than one hundred people, who have been inspired to change their way of thinking. Syen loves adventure and enjoys outdoor sports, such as diving, riding, hiking, and camping. By the age of twenty-six, she had traveled to more than ten countries. Her passions include inspiring others to live with gratitude and living her life to the fullest by sharing her story and lifestyle.

Contact Information
Facebook: https://www.facebook.com/si.yen.14

CHAPTER 30

PULLING YOUR OWN WEIGHT

By Toni Catchings

I remember when I first started playing sports. All my grade school friends were playing soccer, so I wanted to play too. Soccer came easily to me, and I found that I loved the game. As I got older, I became interested in playing for a better team in a more competitive league. The team I was hoping to play with was highly successful. To join the new team, I had to try out. The coaches would have all the aspiring players perform the same tasks, and then they would pick the best performers. I remember watching a tryout so that I could figure out what was required. I realized that, to be successful, I needed to practice, practice, and practice again. My dad would go with me to a local football field and help me practice. Finally, the time for tryouts arrived, and I made the team. I continued to play with this team for the next five years.

The team coach was a little intense. He had really high expectations of his players. For instance, if your uniform shirt came untucked during a game and you didn't fix it immediately, he subbed you out of the game. Practices were always challenging. We started every practice with a one-and-a-half to two miles timed run. Then, we had to complete one hundred sit-ups, followed by jumping rope for a specific number of successful jumps. Once everyone had completed this, then practice began. At the end of practice, we always finished with both fifty and one hundred yard sprints. At this time in my life, I REALLY hated running. But I knew that if I wanted to play with this team, it was just something that I had to do, whether I liked it or not. I was learning how to carry my own weight, although I didn't even know it at the time. I did know that I was definitely part of a TEAM, and that we were working and playing for and with each other both on and off the field.

When I was nineteen, the team was invited to represent the United States in an invitational international soccer tournament in China. Back then, there were no U.S. women's national or Olympic soccer teams. We won most of our games, but we also had a tie with China and a loss to Australia. We still qualified for the semi-finals and made it through to the finals. But, of course, who did we have to play in the finals? Australia again! We had already lost to them once. Did we honestly think we could beat them?

I remember feeling super-nervous when the final game started. We made it to half-time, and the score was still tied nil-nil. I remember seeing my teammates do things that I had never seen them do before during that game. The determination that we had to come back and beat this team was INTENSE! Everyone on the team took care of business. When the final whistle blew, my teammates and I were exhausted by the supreme effort that we'd put into beating our opponents, but the feeling of accomplishment was overwhelming. Final score: USA one, Australia nil.

Now, as an adult, I realize that the coach pushed us so much to make sure that everyone could carry their own weight during a game. We would all be able to push through and continue to play until the whistle blew at the end of the game.

I have another soccer championship memory, where I did not carry my own weight. It was during another final game to determine the best team in the country for our age group. We were tied with the other team, and they had control of the ball. The next play involved the player I was responsible for marking. I remember thinking to myself that she would mess up, and I could relax a little. My poor decision to relax and not give it my full effort changed the game; she made the right play and did it well because I wasn't challenging her. I underestimated her will to win and her ability. Within the next few minutes, that team scored a goal.

I KNEW it was my fault. We never came back from that. The other team won the championship. I had failed my teammates and myself. I remember that poor decision to this day. My lack of effort let down eighteen other people and all our fans. Even now, years later, I often wish that I could go back in time and have that moment back—and make a different decision. Of course, that's impossible, and I must live with it.

Life continued, and I found myself divorced at the age of twenty-seven after five-and-a-half years of marriage. This utterly crushed me. I felt ashamed of the situation because I felt as though I had failed. I was not proud of having to admit that we could not work things out. Now, I had to figure out how to carry my own weight in my new single life. I needed to figure out how to remain living in my house and continue to make a car payment on a small salary. Honestly, I was really struggling with the divorce, both financially and emotionally. I had allowed myself to become emotionally dependent on my husband. Once again, not a proud moment.

I was suicidal during this time, thinking of more than one way to take my life. Of course, my parents and friends knew I wasn't myself, but I never told them about the negative feelings and thoughts going through my head. Then, a friend of mine invited me to join her husband and her at a church service. I went along and, after that, continued to participate in going to church. This was the step I needed to get past the divorce and move on. Allowing God back into my life was a game-changer. That was when I realized that I needed to start carrying my own weight again in my personal life. I made a decision: I needed a better paying job and to start taking care of myself again.

I had always wanted to help people by becoming a paramedic and ride an ambulance. My ex-husband had always discouraged it, but now I went for it. Within months, I applied for a position with the fire department to pursue my dream of being able to help others. I realized that I needed some serious training to prepare for the physical ability test. Once again, I found myself doing the hard work that was needed to be successful. I prepared by pushing (by myself) a full-size pick-up truck through a church parking lot numerous times a week. I also pulled tires loaded with bags of concrete through the park. I was determined to pass the test.

The day of the test arrived. I was extremely nervous but ready for the challenge. I remembered during the last section of the test that I needed to make an adjustment to the equipment, but the voice inside me was telling me, "Don't stop! Keep going, just do not stop!" I listened to that voice and completed the test with only one second left on the timer. The test wasn't perfect, but I was successful because I had prepared for the day.

Welcome to the Fire Department! Now I had to figure out what was required to carry my own weight daily in rookie school as a firefighter and as a paramedic. All rookies must prove themselves once they graduate and are assigned to a specific station. This is a situation where you have to pull your own weight, or you might as well go and find another job. Understandably so; firefighters have to go into burning buildings when everyone else is running out, and they may need to carry someone out. Even for the guys, proving yourself as a trustworthy member of the station crew is hard. But for a female, Holy Cow! Even the general public doesn't believe that a woman can be physically strong enough for this career. The answer to proving yourself is to take things one step at a time, and always give one hundred percent effort to every task. Even the menial tasks, like mopping the floor and cleaning the toilets, you must give it your all. When a fire call comes in, you must step up and be prepared to use everything that you were taught in rookie school. On the inside, your adrenaline is pumping like crazy, but on the outside, you appear calm and collected. You have to keep moving forward and taking one step at a time, even when fear is setting in.

Over the last twenty-seven years as a firefighter and paramedic, I have experienced a lot of different situations. I reflect on each one to determine whether I pulled my own weight. I always ask myself if I could have done anything differently to affect a different, better outcome? One specific incident that I vividly remember was when a thirty-year-old male was shot multiple times by teens who were just out shooting randomly at people for no reason. The fire engine arrived on the scene first, and the man was still talking and able to communicate. My partner and I arrived in the ambulance a couple of minutes later. Now, the guy was no longer talking. He was unconscious. We placed him in the ambulance and began taking care of him. But by the time we arrived at the hospital, he had deteriorated. We performed CPR but were not able to save him. The immediate feeling I had was that I had failed. I felt I had let the man and his family down. After all, I am supposed to be the angel that comes in and makes a positive difference.

This incident replayed in my mind for days. Every time I replayed it, I knew there was nothing else I could have done. Both my partner and I had pulled our own weight. We knew what to do, how to do it, and when to do it. We were prepared for the situation. We made all the right decisions and had both given one hundred percent effort to saving the man, but we still had no control over the final outcome.

Of course, it's at times like these when we have to accept that we're only human, there's only so much you can do in some situations, and despite doing your utmost, you're going to be unsuccessful at times. I have failed and am going to fail again at times, just as we all do. But can something truly be called a failure if I give one hundred percent of myself in the attempt? Yes, I may be unsuccessful, and I get very disappointed and unhappy when things don't work out the way I want, yet I find that it is easier to accept the outcome if I know I gave it my 100%.

Stuff happens to everyone. That's life. But you have to decide if you want to be an asset or a liability. You have to ask yourself if you're going to be crippled by what happens to you, or empowered by it. I choose to be empowered! I want to carry my own weight in life and continue to get up every day and improve on my life. I realize that the real struggle is within myself, as it is within all of us, and the real battlefield is in my own mind. To achieve what I want to achieve, I must relentlessly battle against the negative thoughts, the lack of energy, the bad days that come to us all. I understand that the decisions I make in my mind every day affect my everyday actions and outcomes. Do I go to the gym today? Or do I take it easy and chill? Do I practice today? Or do I decide that it will be okay if I don't prepare?

To me, it's an incontrovertible truth that we control our own destinies through the decisions, or lack of decisions, that we make numerous times a day. All of us know this, but often we still let life make the decisions for us, and we simply go along with them, even if they're not what we really want. I believe that we must do the opposite and make the decisions that determine our lives for ourselves. Are you happy to let circumstances control you? Or do you want to control the circumstances? Life seems to me like a chain of events requiring decisions in every corner, and a wrong decision or a failure to make a decision is like a weak link in that chain that can affect our whole lives.

I say, do not allow yourself to be caught up by circumstances—don't be that weak link. Instead, take some time to figure out what needs to change to make you a stronger link—and then take the appropriate action! Keep taking one step at a time and moving forward. Try to keep improving yourself, and always keep learning. It's up to us as individuals to decide what needs to happen to make sure that we pull our own weight in our lives.

BIOGRAPHY

Toni Catchings is best known for two things: Being a strong athletic woman and her compassion to help others. She was inducted into the Sports Hall of Fame at Texas A&M University-Commerce in September 2015. She continues to serve as a firefighter and paramedic. She has been promoted through the ranks over the last twenty-seven years and currently holds the rank of Captain. She enjoys helping young girls realize their true potential and to believe in themselves. Toni is also the owner of Catch 22 Home Solutions.

Contact Information
Facebook: http://facebook.com/toni.catchings.3
Instagram: https://www.instagram.com/monkee2022/

CHAPTER 31

CHOOSE MENTORS AND APPRECIATE FAILURES

By Tonika Bruce

I believe I was destined for greatness with a purpose to lead people. My greatest satisfaction has always come from helping people. I can look at a process and quickly suggest ways to improve it, and think of several related businesses, only to see that idea become a multi-million-dollar business within three or four years.

When 9/11 happened, I was 23-years-old. I thought it was the worst day of my life. My business seemed doomed. No one was buying or selling; people were too busy, in shock, glued to the TV. I went from making 30K a week to being afraid to buy a bag of ramen noodles. I lost everything—money, pride, and the fire within. I didn't know what to do. I had no real friends. I couldn't ask my family for help. How could I ask anyone who hadn't ever been in such a situation for help?

I credit this as my turning point—the moment when I realized that chasing money isn't the right way to do things.

I kept asking myself the same questions over and over: How did this happen to me when all I know is how to be successful? Do I see success only one way? In just one area of my life? With only certain people?

At that age, I didn't take the time to reflect on what had gone wrong; I was broke, hungry, and prideful. I thought God was angry with me. My best friend had committed suicide, and I knew how that affected me and everyone else, so that was NOT an option. It seemed as if my only choices were to hustle or just give up.

I graduated from college, enjoyed a successful career playing basketball, and was ready to continue living the so-called all-American dream. I reconnected with the love of my life, got a great, high-end apartment, and started working until the time was right to go back to school. But even with three jobs, money was tight.

Then, I saw an ad that said $20/hour: I called the number and got an interview. I arrived early and was immediately drawn to Alan, the presenter. Watching him completely changed my life. I was deeply impressed by the way he commanded the room and connected easily with the 50 people in the room. That day, I felt I'd seen what success looked like—wrapped in an expensive suit, with conviction in his voice, Alan had the ability to make people *feel* what he felt. I took the job and used my rent money to purchase the required kit. It was my first direct sales position; even with zero sales background, my competitiveness, my desire to be the best, and make a ton of money drove me from the start.

My biggest growth came from modeling Alan and reading several of the books he referenced.

Being naturally gifted and ambitious, I wanted to be the best; I advanced quickly and earned the opportunity to work with the owner of the company in Ft. Lauderdale. One of the most important lessons I learned is that having the wrong mentor can negatively influence your chances of success if you aren't strong enough to fight against their effects. After a few months of working in Ft. Lauderdale, I was ready to go out and open my own location in Dallas, TX. I prepared by attending events with people like Tony Robbins, Mark Cuban, Zig Ziglar, and Les Brown. I learned that everything was possible—when I took action.

My plan was to open 300 locations in 10 years. We opened eight offices in 6 months. I did interviews, hiring and training over 3000 sales representatives. At the young age of 23, I was easily making a six-figure income monthly. I studied, developing myself into a great speaker and teacher, even at events outside of my own company. And then, it happened: I didn't just fall, I plummeted and landed flat on my face.

After that day, I decided it would be best if I didn't try again. I was broke—I had maybe $50 at best, plus credit collectors were hounding me with daily calls. Never in my life had I ever felt so scared, hurt, or angry, and I had no idea of what to do to get out of the situation.

It wasn't long before that $50 became $10, which was when logic and necessity won out. I desperately needed money, so I got a job to fit in better with the people around me. First off, I tried restaurant management. True, the hours were long, and I smelled of cooking every day, but the money was great, and I soon became successful again. I even managed to grow our restaurant to new levels. My time in restaurant management taught me the value of giving good service, as well as humility, but soon it felt like it was the right time to leave the fajitas behind and try something closer to my heart.

That's when I first got into basketball coaching. I had the opportunity to watch the grind, hustle, and enthusiasm in my good friend, Dorian. He commanded attention, and his intensity was infectious. Before long, his positivity drew my thoughts back to the dreams I'd once had of greatness. My purpose and passion for leading, training, and developing people was leading me back to my roots, and my first passion—basketball. Being naturally competitive, I loved winning, so playing for championships was always top on my list of great ways to spend time. In a short period, I'd exchanged my expensive suit for gym shorts and sneakers. Okay, so the "vehicle" had changed, but so had the quality of the mentor. That, as I was to discover, was the missing piece of the puzzle.

There's no denying that I enjoyed incredible success as a basketball coach. But, although I coached a team that won five championships in a single season, it seemed as though the athletic director's sole mission in life was sabotaging my players and team—just because we were successful! It was this that helped me officially confirm my understanding of mentors and the influence they can have upon you. It made me realize how important it is in life to carefully choose the people and things you learn from. As I discovered, if you surround yourself with the wrong ones, your greatest accomplishments can be spoiled as a result.

It's sometimes hard to accept, I know, but trust me when I say that a bad mentor can come in many unexpected forms; it can be a boss, the company you work for, a friend, co-worker, or parent. At the time, I couldn't help wonder if I was the only one who had come across a boss or a group of people apparently dedicated to causing me to fail. I'd always believed that bosses, managers, and the people around you are supposed to be rooting for you and helping you to succeed. Clearly, I was wrong.

Then, without warning, things for me just seemed to hit a wall; I was at the top and climbing in my coaching career. I had over ten interviews and got seven job offers, just not the three I wanted. I felt like a failure again. I felt like I was deliberately being brought low, and this time I wasn't sure I had the will to get back up.

After my grandmother passed away, I was led into a completely new and different field—nursing. I had gone from sales, public speaking, coaching teenagers to wanting to care for geriatric and cardiac patients. But I wasn't sure that nursing was even worth getting excited about. Maybe this was God's way of saying that he wasn't mad at me anymore and that he was giving me another shot? One thing I was sure of, though—this was significant. It told me that my life meant something, and whomever I was becoming, it better be good! By this time, I'd witnessed first-hand the terrible standard of care that patients were receiving and wanted to remedy that by being the opposite—I wanted to be THE BEST!

Nursing school was absolutely exhausting! Anyone making it out alive has rightfully earned the title WARRIOR BEAST. The small town of El Dorado, Kansas, hosts a hidden gem, one of the best nursing schools on the planet. During this time, more than ever, my growth as a leader was challenged and elevated. Yes, there was a moment when I felt like quitting, when hearing the voice in my head of my high school coach, Coach P, saying, "Finish what you start and make things better than you found them," just wasn't enough. My turning point came in the third semester, when I came across an instructor named Sherry. Sherry was super tough, which made me want to be in her clinical class. Not only was Sherry tough, but she also demanded a lot more from her students than other instructors. Under Sherry's teaching and mentorship, my previous expectations of what was "standard" shot up the scale. She made me see that what I'd been giving and contributing so far was simply not good enough.

In nursing school, I became president of my class for two consecutive terms in a row and formed lasting friendships with several classmates and instructors. I was honored to be the commencement speaker at our graduation ceremony. This re-ignited my fire for public speaking and encouraged me to study for my master's degree in Executive Leadership.

My quest to give my patients the best care possible was often clouded and compromised by a lack of resources, insufficient staff, and lack of teamwork. I took a job to be a part of something great and believed so deeply in the hiring manager's vision that I was willing to drive 90 minutes each day to reach the goal I had in mind. I wanted to be a part of that success, but when both my managers left, my motivation changed, and I found myself once again no longer positioned for success.

Arriving at a place where success was truly represented in the people around me, and finding a suitable mentor were now the most important things to me. I wanted to experience success, not just read about it on a vision board, so I continued to search for my "fit." My purpose changed from wanting to lead people to searching for a leader to lead me in the right direction.

The day finally came when I decided that I was going to be the best leader I knew and that, to achieve that goal, I had to make myself better. By chance, I crossed paths with one of the most influential leaders and mentors I have ever come across in nursing—my then Chief Nursing Officer, Molly. Molly's motto was, "make it happen." Her smile made her look like she was having so much fun, plus she had the passion that motivated her to work to elevate our team. I saw success ahead of me again. This time, however, the suit was accompanied by stiletto heels and flip curls. I was drawn to Molly's drive, passion, and humility. I wanted to be like her.

I had a patient who'd had surgery on both eyes and couldn't see my face. I attended to her in recovery, and she said I was just the person she needed to be her nurse. She asked my name, and when I said my first name only, she asked for my last name. When I told her, she smiled and said, "Ah, Dr. Bruce!" She then proceeded to tell me that I wasn't going to be just a nurse for long. I was going to speak and lead people while heading up a successful business that would grow internationally. She said that I might lose a few people along the way, but that everyone who jumped and stayed on board was going to catch that same fire that burned inside me. She spoke of the things I knew but never shared with anyone. The interesting part was that I had just graduated with my master's degree and had just been approved for doctoral admissions.

Shortly afterward, I was injured at work. The injury was severe, causing a neuro deficit in my leg and foot. Most of the time, I couldn't feel my toes. I couldn't walk, bend over, or sit down, let alone work. As a result, my income plummeted from six figures to around the average household income. It seemed to me that I'd failed again, and, this time, I was completely powerless to take action to change my situation. I had no choice but to be still. Initially, I was angry that this had happened to me, until I suddenly realized that I was actually prepared for the situation and that, in fact, it had happened for me.

The injury slowed me down, gave me time to reflect on just how I'd arrived at that point, what my purpose in life was, and how I could turn my situation into a success. Looking back, I saw that I'd gone through just about every circumstance to prepare me for this challenge.

Around this time, I was shown an opportunity involving network marketing. Without a clear plan, I accepted it and figured I had to act this time. I'd done enough self-reflection to realize that the job I'd valued so much didn't value me at all. I decided it was time to get back to doing things that fulfilled my purpose.

Starting my own network marketing business turned out to be the best decision I ever made in my life. I had all the necessary skills, desire, motivation, and fire already in me to make a go of it. And, yes, that patient was correct: Some people, including family members and friends, refused to come along on the ride. But those that did, I'm happy to say, are sharing the fruits of success right alongside me. For me, helping others reach their financial potential, especially during these tough times, continues to be deeply personally rewarding. I often think about that patient: She was right about my business being a success. One day, I hope to give her a big hug and treat her to a nice lunch.

At this point in my long and sometimes challenging journey through life, I've come to realize that no matter what my job was or how many times I failed in the past, all I needed was to see that my success had no boundaries and that, sometimes, we just need to be shown what success actually looks like.

BIOGRAPHY

Tonika Bruce is a registered nurse, speaker, change agent, and serial entrepreneur with over 20 years of experience in building businesses and teams. She is a 6-figure income earner and shares an insatiable desire for helping others find success, purpose, and prosperity. She relishes living a life of a higher purpose and is the founder of multiple non-profit organizations. Tonika strives to amalgamate her love for serving people and expertise in business to help others unlock doors to financial freedom. Tonika found success in multiple niches, including nursing, entrepreneurship, business and basketball coaching, and executive leadership.

Contact Information
Facebook: https://www.facebook.com/tonikabruce11
Instagram: http://instagram.com/tonikabruce
Website: http://tonikabruce.com

CHAPTER 32

DREAMS ARE FOR SUCKERS

By Whitney Tello

New mom. Divorced. Unemployed. I did not plan for it in my life planner. This was not the life I wanted, but this was, in fact, the life I was living. The *unimaginable* was now my *reality*. Fate played a cruel joke on me, and within eighteen months, my entire life changed its course.

Several months after my son was born, my marriage of five years began to fall apart. Soon after, things spiraled rapidly out of control and I was left with the hardest decision of my life—be a single mother or stay in an unhealthy relationship. Even though the force of habit lured me to stay, I knew it wasn't right. I was worth more. If I had stayed in that same situation, I would have settled for much less than I deserved.

I spent sleepless nights, frustrated, why could he not see my love for him? *Why could he not see what he had right in front of his face?* Such thoughts occupied my mind. He had a wife that adored him and a 'brand-new' perfectly healthy baby boy. This was supposed to be *our* time. We spent two of our five years of marriage dealing with the dreaded pains of infertility. Our son was supposed to be our beacon of light, our reward for all our pain and suffering. Little did I know, my suffering through the years of infertility was minuscule when confronted with the devastation of a broken marriage.

Imagine two people in a rowboat with stationary oars, both of them responsible for rowing their oars. But if they do not row in harmony, that boat is not going anywhere anytime soon. Much like that boat, my marriage was moving in circles with no clear destination in sight. After endlessly rowing alone, my arms refused to budge. I filed for divorce. I had a surprise in store for me. Two weeks after my divorce finalized, I was let go from my job—a job in which I was recently promoted and was on track to achieve much more.

I was in the boxing ring of life, 12 rounds in and exhausted. Life was throwing all its best punches at me. I was knocked down but not yet knocked out. I had a choice: do I continue onto the 13th round, or do I give up? I could faintly hear the referee counting to 10 as my inner monologue kicked in.

"Why is this happening to me?"
"What else could go wrong?"
"This isn't fair."
"I didn't sign up for this."
"Why can't I catch a break?"
"What am I going to do?"
"You'll figure it out."
"Can he stop counting already and let me think?"
"*Get up.*"

I got up. Shaken and confused, but I carried myself into the 13th round and beyond. I realized that life takes unexpected turns. The choices of others, both indirectly and directly, affect you. Trauma and loss are intricately woven in the circle of life. At first, I focused on *why*. But those questions led me nowhere, except down a rabbit hole of confusion and dismay. I was asking the wrong questions; I needed to ask *how*.

"How will I get out of this situation?"
"How will I provide for my son as a single mother?"
"How will I show my son that he has a strong and resilient mother?"
"How will I create more time and achieve financial freedom?"
"How will I ensure that my financial future is in my own hands?"
"How do I wish to be remembered?"

"How will I pass on my legacy to my son if I don't start it *now*?"

Although I never desired to be an unemployed, single mother, and the circumstances were wildly out of my control, there was *one* thing that was still under my control—my actions. What I did about it was 100% my responsibility. I got up before the referee got to 10. I put my future back in my own hands. I regained control of my life and put myself in the driver's seat again.

Accepting your reality does not mean understanding your reality; there is a difference. Merely possessing knowledge doesn't produce results. People often talk about how they know what they should be doing, but they aren't doing it. So, I threw in the towel in trying to make sense of my new life. I leaned into accepting rather than understanding, and changes started to happen, slowly but surely. I made a decision that *why* no longer mattered, but *what* I am going to do about it, does.

Education doesn't prepare you for life. Only life can prepare you for life.

Are You a Sucker?

Have you ever been suckered, much like me, at a young age, to buy into the mentality that you should follow your dreams? To think that you should always believe in yourself? Or even believe that you can be anything you want to be in this world? Did you also believe that if we go to school, get into a good college, get good grades, land a good job, get married, and have kids, life would be all fine and dandy? So why is it I was a divorced, single mother, and unemployed even with a degree in business? I checked all the "right" boxes exactly how society taught me, but life still happened.

Once we transition into adults, 'dreaming big' is criticized, dismissed, rejected, and even mocked. "Aim for the stars" fades into the stern noise of "be grateful for what you have." Is it wrong to want more? Be more? Do more? It wasn't wrong for a child, then why so for an adult?

We've all heard the phrase money doesn't buy happiness. That is true. However, we also fail to mention that money does provide the means to create the lifestyle you desire for your own level of happiness. Does gardening bring you happiness? Money enables you to pay for the soil, plants, or gardening tools. Does traveling with your family bring you happiness? How do you pay for the gas, flights, accommodations, food, etc.?

Once I purged myself of self-sabotaging thoughts that I do not deserve more, I went for it at full speed. I have now found a life full of happiness and purpose. My mantra still rings true now as it did at the beginning of the 13th round: I don't want money to be rich, I want money to *enrich*. Yes, you can do both—have money and still enrich the lives of others.

I Will Ask Again, Are You a Sucker Too?

Have you accepted your reality? Are you ready to be in the driver's seat? You create the life you want. You determine your future. It's up to you to bring your vision to life.

If you have a vision, do you see it clear as day? Do you live it and speak it into existence? Good, then you're a sucker too. Welcome to the club! You got that vision, right? Everything in life, including your success, depends on you *protecting your vision*. You are the gatekeeper of your vision. The only person truly stopping you from living your dream life is you.

THE END

BIOGRAPHY

Whitney Tello is a self-made entrepreneur who fought her way to the top when life knocked her down. From becoming a first-time mother to a divorcee to an unemployed single mother, Whitney was still determined to leave her mark on this world. Growing up thirty minutes outside of New York City, Whitney was accustomed to dreaming big. With a degree in Business and Spanish, and digital marketing expertise, she jumped two feet in when it came to starting a career in network marketing. Now through years of self-development and self-discovery, Whitney has made it her mission to help others get unstuck in life by sticking to their dreams and transforming them into realities. Some of her many passions include spending quality time with her son, fitness, travel, and never giving up on her lifelong quest to master the art of making the perfect pancake.

Contact Information
Facebook: https://www.facebook.com/13TeenthRound/
Instagram: https://www.instagram.com/whittywits/
Website: https://www.13teenthround.com/regain-control

www.ingramcontent.com/pod-product-compliance
Lightning Source LLC
Chambersburg PA
CBHW031918240526
45464CB00021B/229